REPORT

T0302506

Best Practices in Supplier Relationship Management and Their Early Implementation in the Air Force Materiel Command

Mary E. Chenoweth, Nancy Y. Moore, Amy G. Cox,
Judith D. Mele, Jerry M. Sollinger

Prepared for the United States Air Force

Approved for public release; distribution unlimited

RAND PROJECT AIR FORCE

The research described in this report was sponsored by the United States Air Force under Contract FA7014-06-C-0001. Further information may be obtained from the Strategic Planning Division, Directorate of Plans, Hq USAF.

Library of Congress Cataloging-in-Publication Data

Best practices in supplier relationship management and their early implementation in the Air Force Materiel Command / Mary E. Chenoweth ... [et al.].
p. cm.
Includes bibliographical references.
ISBN 978-0-8330-5144-8 (pbk. : alk. paper)
1. United States. Air Force Materiel Command. 2. United States. Air Force—Procurement—Evaluation.
3. United States. Air Force—Supplies and stores—Evaluation. I. Chenoweth, Mary E.

UG1123.B47 2012
358.4'16212—dc22

2011014741

The RAND Corporation is a nonprofit institution that helps improve policy and decisionmaking through research and analysis. RAND's publications do not necessarily reflect the opinions of its research clients and sponsors.

Published 2012 by the RAND Corporation
1776 Main Street, P.O. Box 2138, Santa Monica, CA 90407-2138
1200 South Hayes Street, Arlington, VA 22202-5050
4570 Fifth Avenue, Suite 600, Pittsburgh, PA 15213-2665
RAND URL: http://www.rand.org/
To order RAND documents or to obtain additional information, contact
Distribution Services: Telephone: (310) 451-7002;
Fax: (310) 451-6915; Email: order@rand.org

Preface

The Air Force is under great pressure to reduce the cost of its logistics operations and to improve their performance. The Air Force hopes to use the realized savings to fund other priority programs, such as force modernization, recapitalization, and carrying out operations. Commercial firms have had success in reducing their costs by using a technique called Supplier Relationship Management (SRM), which uses intensive management of supplier relationships to reduce costs for both the buyer and supplier and increase supplier performance. Although the Air Force has implemented an SRM program, concern over the pace and success of its supplier-development efforts led the Air Force to ask RAND Project AIR FORCE (PAF) to identify how the Air Force might enhance its SRM effort to reduce total costs more rapidly.

The research reported here should interest those involved in the Air Force's logistics transformation efforts and other efforts applying business practices to government-run activities.

The research described in this report was performed for a fiscal year (FY) 2007 study titled "Strategic Management of Maintenance." The work was sponsored by the Director of Transformation, Deputy Chief of Staff for Logistics, Installations and Mission Support, Headquarters United States Air Force (AF/A4I), and the Deputy Assistant Secretary for Contracting, Headquarters United States Air Force (SAF/AQC). The study was conducted within PAF's Resource Management Program.

RAND Project AIR FORCE

RAND Project AIR FORCE (PAF), a division of the RAND Corporation, is the U.S. Air Force's federally funded research and development center for studies and analyses. PAF provides the Air Force with independent analyses of policy alternatives affecting the development, employment, combat readiness, and support of current and future air, space, and cyber forces. Research is conducted in four programs: Force Modernization and Employment; Manpower, Personnel, and Training; Resource Management; and Strategy and Doctrine.

Additional information about PAF is available on our website:
http://www.rand.org/paf/

Contents

Figures

Tables

Summary

Background, Purpose, and Approach

As is the case with all military services, the U.S. Air Force (USAF) is under pressure to reduce the costs of its logistics operations while simultaneously improving their performance. In FY 2008, 43 percent of the total Air Force budget went to purchases of goods and services—including all weapon system support costs, base operating support, military housing and construction, services, activities outsourced to suppliers through A-76 competitions, and upgrades managed by system program managers at the air logistics center (ALC) level—from the private sector. Thus, any efforts to generate significant savings must target purchases from the private sector.

Since 2002, as part of its Purchasing and Supply Chain Management (PSCM) initiative, the Air Force has had a program called Supplier Relationship Management (SRM) within the Air Force Materiel Command (AFMC) that has the goal of reducing the cost and improving the quality and performance of purchased goods and services by carefully managing relations with suppliers to the mutual advantage of each. Commercial firms have long had such programs and have garnered large savings from them. However, the scope and pace of the savings generated by the SRM program at AFMC have been less than anticipated by senior leaders. As a result, the Air Force asked PAF to identify steps the Air Force can take to improve the outcomes of its SRM program, particularly for contracts for sustainment spares, repairs, and other depot-level services.

In performing this research, we first reviewed the academic, business, and defense literature on best SRM practices and interviewed representatives of commercial best-in-class firms that have successfully implemented SRM to identify the most effective practices. We then assessed how the Air Force had implemented SRM, to determine whether it was employing the best business practices in its SRM program and to uncover practices or policies that might be making the implementation of SRM less effective than it could be. This step included an analysis of annual spending with key SRM contractors as well as intensive interviews with Air Force personnel. Our final step was to recommend actions that the Air Force could take to improve its implementation of SRM.

Findings

Our review of the literature and our interviews with representatives of commercial firms that have successful SRM programs show that the six practices listed below characterize best SRM practices:

- Manage total business with each supplier (consolidate contracts, tie future business to performance).
- Measure and shape supplier performance (establish performance measurement system, rank suppliers, set targets, reward performance).
- Involve key suppliers early in product design (leverage their design capabilities and knowledge of manufacturability and innovation, reduce complexity).
- Host high-level meetings that promote dialogue with suppliers (demonstrate mutual commitment to the relationship, promote dialogue on expectations and ways to improve, share future plans and technology roadmaps, present awards).
- Recruit skilled personnel (recruit experienced personnel who have the right qualitative and quantitative skills).
- Develop personnel so they have a thorough knowledge of suppliers (educate personnel so they know suppliers' processes, costs, capacities, and capabilities and can work with and help suppliers fix processes to meet current needs and continually improve).

The Air Force is already implementing several of these best practices. Focusing primarily on National Stock Number–level (NSN-level) spending of the eight commodity councils (CCs) in the early days of the SRM initiative,[1] AFMC identified top suppliers by dollars and assigned them to SRM teams where the locus of SRM business took place. Supplier scorecards were developed to show how suppliers were performing overall and in detail. Supplier summits have been held since 2004 that bring together both Air Force and supplier senior leadership to talk about goals and objectives and act as a clearinghouse for solutions to various problems. Joint Improvement Initiatives that are mutually agreed on by the SRM teams and suppliers have formed the basis of developing improvements in processes and outcomes. SRM team members are usually full-time employees who have been selected on the basis of their skills for self-initiation, powers of persuasion, and hard work.

However, several major factors seem to be keeping the Air Force's SRM program from accomplishing more. First, the supplier business base to which SRM has been applied is small and minimizes incentives for the largest SRM suppliers to devote time and resources toward eliminating waste and increasing efficiencies in their processes and organizations. Second, well-defined goals do not appear to be communicated clearly and consistently among SRM teams or within AFMC in general. We conclude that the result of not having well-defined goals is that many SRM staff do not appear to understand their responsibilities clearly, and other internal stakeholders do not appear to understand the purpose of SRM; this can slow the implementation of SRM concepts in new contracts. Third, SRM teams do not appear to have sufficient numbers of personnel to manage the number of suppliers and size of spend at most sites. Fourth, the program lacks sustained senior leader support needed to institutionalize SRM. Finally, the way in which SRM integrates into other AFMC initiatives creates organizational complexity that hinders effective and efficient implementation. The commitment of SRM personnel to their job may erode without strong leadership from the Strategic Supplier Relationship manager (SSRM) who leads the SRM team, evidence of the benefits of SRM, means of bringing in stakeholders early on and gaining their support, and support to reward

[1] The purchasing of spares and depot-level contract repairs has been centralized to eight CCs located at the three ALCs. The eight CCs are Aircraft Accessories, Aircraft Engines, Aircraft Structural, Communications-Electronics, Instruments, Landing Gear, Secondary Power Systems, and Support Equipment.

and maintain SRM team members' career field development. Addressing these factors could energize SRM efforts by giving them greater influence and direction.

Recommendations

Our analysis suggests the following recommendations for the Air Force to pursue to better realize the goals of SRM:

- Expand the business base.
- Institutionalize SRM.
- Develop analytic capabilities.

Expand the Business Base

The current business base of SRM at AFMC is primarily restricted to weapon system spares and repair purchases. These represent at most 10 percent of the Air Force's direct spending budget and only a fraction of the business base that some of the largest SRM suppliers have with the Air Force. Successful commercial firms integrate SRM concepts across their business units to include all of the business that a supplier has with the company. The relatively small portion of the business base currently considered within SRM in the Air Force significantly limits the scope of the savings that could be achieved.

Expanding SRM to include a greater portion of Air Force spending with suppliers can increase their incentives to improve. Such an expansion would include all spending within the ALCs and spending at the product centers, where new weapon systems are acquired. One means of linking the sustainment business base (the little "a") to the acquisition business base (the big "A") would be to develop an integrated supplier scorecard for the whole of AFMC. Another means might be for the product centers to develop their own SRM initiative to operate above the program level—according to whatever the law permits—and act as a clearinghouse for identifying common issues that, if addressed by the supplier in one program, could benefit those in others. Still another means might be to establish a joint product center–ALC SRM program where the total business base of key suppliers is included, with both types of businesses (big "A" and little "a") represented.

Institutionalize SRM

Institutionalizing SRM to make it an integral part of managing key suppliers would provide it more influence and direction. Steps toward institutionalizing SRM include the following:

- sustained support from top management, which would communicate that top management values SRM and intends to cascade those values throughout the enterprise and supply chain
- formal SRM policies and guidance that clarify objectives, roles, and responsibilities of SRM teams and key stakeholders
- processes, official recognition, and assignment of the duties of the SSRMs
- career incentives and development programs for SRM team members
- formal training programs
- cross-functional support available to SRM teams

- process changes that facilitate implementation of SRM initiatives across organizations and locations
- methods of measuring and sharing metrics of supplier performance improvements across the enterprise.

Unless such steps are taken and unless SRM can demonstrate real improvements, SRM risks becoming marginalized and losing resources, importance, and, inevitably, influence. Institutionalizing SRM provides a means for AFMC to work directly with its key suppliers across the total Air Force business base to improve those things on the customer and supplier sides that unnecessarily keep costs high and affect quality and performance.

Develop Analytic Capabilities

AFMC should continue to grow an analytic capability that can develop and refine supplier scorecards; identify and justify targets of opportunity for improvements; and develop business cases needed to secure support from top Air Force management, various stakeholders, suppliers, and Defense Logistics Agency and organic maintenance. Implementation of such an analytic capability for SRM should benefit directly from the Air Force's new enterprise resource planning system, the Expeditionary Combat Support System (ECSS), which should facilitate the collection of data needed for SRM-related analyses.

SRM teams will also need to keep pace with best practices. The companies we interviewed have made this capability organic to the team. This capability becomes especially important to the SRM team's ability to demonstrate the benefits of SRM to key players, most notably the supply chain managers.

Conclusion

SRM has grown in importance in the private sector over recent years. AFMC's experience with SRM suggests that while it has tried to implement many of the best practices, some obstacles prevent it from reaping real benefits from the initiative. If these obstacles are addressed, SRM could become a significant strategic tool for AFMC in seeking real cost and performance improvements with its key suppliers.

Acknowledgments

We first wish to thank our project sponsors, Grover Dunn, Director of Transformation, Deputy Chief of Staff for Logistics, Installations and Mission Support (AF/A4I), and Charlie Williams Jr., then Deputy Assistant Secretary for Contracting, Office of the Assistant Secretary of the Air Force for Acquisition (SAF/AQC), and now Defense Contract Management Agency Director (DCMA-D). They provided the motivation for this research. Next, we would like to acknowledge the many individuals who provided us assistance with obtaining data or information and those who gave generously of their time to be interviewed. These individuals are listed with the rank and office affiliation current at the time of the research.

AFMC/PKT (Analysis and Program Support Division, AFMC Headquarters Contracting) provided data on information to help us identify those classes of items managed by CCs and codes to identify private-sector entities associated with their 21 key suppliers. These were used in identifying spend for SRM key suppliers.

We would like to thank those individuals who were SSRMs at the time of this study, some of whom were interviewed one or more times. In alphabetical order and their offices at the time of the study, they include Roger S. Correll, Director of Contracting, Ogden Air Logistics Center (OO-ALC/PK); Lorna Estep, Directorate of Logistics, Air Force Materiel Command (AFMC/A4); Patsy Reeves, Director of Contracting, Warner Robins Air Logistics Center (WR-ALC/PK); Bonnie Taylor, Director of Contracting, Oklahoma City Air Logistics Center (OC-ALC/PK); and Thomas Wells, Director of Contracting, Air Force Materiel Command (AFMC/PK). These individuals gave us full access to their SRM leaders and personnel. We interviewed their team members, both in groups and singly. Though we promised anonymity, we thank each of these individuals and appreciate the information and reflections they shared with us.

We thank Steven Martinez, DCMA-OCM, who explained DCMA's role with respect to supplier relationship management, and Scottie Knot, DLA/J-3 (Logistics Operations), who described how DLA is developing its major suppliers.

We thank Col Sid Banks, Air Force Global Logistics Support Center (AFGLSC)/448th Supply Chain Manager, Wing Vice Director (SCMW/CL), and Gregory May, AFGLSC/448th Supply Chain Manager, Group Manager (SCMG/CL/DD), for their generous time and assistance in explaining where SRM fits within the Global Logistics Support Center (GLSC) organization, what has been decided, and what is still in the process of being decided. We have cited this information in the text and references.

We would also like to thank others whom we briefed; each one provided comments and feedback that we responded to and incorporated. We thank Michael Aimone, Assistant Deputy Chief of Staff for Logistics, Installations and Mission Support, Headquarters U.S. Air Force (AF/A4/7); Maj Gen Polly Peyer, Director of Resource Integration, Office of the Deputy

Chief of Staff for Logistics, Installations and Mission Support (AF/A4/7P); Thomas Wells; Col Lee K. Levy, AFMC/A4; Col Brent H. Baker, Sr., Commander, Air Force Global Logistics Support Center Provisional (AFGLSC/CC); and Col Edward J. Koslow, Air Force Materiel Command, Chief of Staff (AFMC/CS).

RAND colleague Carol Fan and Lisa Ellram, Rees Distinguished Professor of Distribution, Farmer School of Business, Miami University, Oxford, Ohio, provided thoughtful and thorough comments that helped refine and improve earlier drafts of the report. We also thank Laura Baldwin, who provided many helpful comments that improved this report.

Abbreviations

AFGLSC	Air Force Global Logistics Support Center
AFMC	Air Force Materiel Command
AFMC/A4	Directorate of Logistics, Air Force Materiel Command
AFMC/PK	Directorate of Contracting, Air Force Materiel Command
ALC	air logistics center
ASG	aircraft sustainment group
ASW	aircraft sustainment wing
BP	a Honda continuous improvement process called Best Position, Best Productivity, Best Product, Best Price, and Best Partners
BRAC	Base Realignment and Closure
CAGE	Commercial and Government Entity
CAMP	commodity acquisition management proposal
CC	commodity council
CLS	contract logistics support
CSW	combat sustainment wing
DCMA	Defense Contract Management Agency
DLA	Defense Logistics Agency
DoD	Department of Defense
DUNS	Data Universal Numbering System
eLog21	Expeditionary Logistics of the 21st Century
FAR	Federal Acquisition Regulation
FPDS-NG	Federal Procurement Data System–Next Generation
FSC	Federal Supply Class
FY	fiscal year
GCSS-AF	Global Combat Support System–Air Force
GE	General Electric
GS	General Schedule

GST	Garden State Training
HAM	Honda of America
HP	Hewlett-Packard
JII	Joint Improvement Initiative
MAJCOM	major command
MICAP	mission-capable
NAPM	National Association of Purchasing Management
NSN	National Stock Number
OC-ALC	Oklahoma City Air Logistics Center
OEM	original equipment manufacturer
OO-ALC	Ogden Air Logistics Center
PAF	RAND Project AIR FORCE
PBL	performance-based logistics
PPI	Producer Price Index
PPM	parts per million
PSCM	Purchasing and Supply Chain Management
R&D	research and development
SCM	supply chain manager
SCMG	supply chain management, group
SES	Senior Executive Service
SIMP	Supplier Initiative Management Plan
SPC	Statistical Process Control
SPM	system program manager
SPO	system program office(r)
SRM	Supplier Relationship Management
SSD	supplier strategy document
SSM	system support manager
SSRM	Strategic Supplier Relationship manager
STARS	Supplier Tracking and Rating System
TCO	total cost of ownership
TQM	Total Quality Management
USAF	U.S. Air Force
USTRANSCOM	U.S. Transportation Command
WR-ALC	Warner Robins Air Logistics Center
WS SCM	weapon system supply chain manager

CHAPTER ONE

Introduction

Background

The Air Force finds itself under great pressure to cut the cost of its logistics activities while improving the effectiveness of their performance to make budgetary room for the highest priorities of the Air Force, including force modernization, recapitalization, and carrying out combat operations. These budget pressures will likely continue, particularly as supplemental funding declines.

The Air Force budget has three major components: personnel, weapon system acquisition, and purchased goods and services.[1] The Air Force has made some progress in reducing the costs associated with these activities. Cost-cutting and streamlining measures, such as the closure of two air logistics centers (ALCs) and multiple bases as the result of Base Realignment and Closure (BRAC) decisions, have reduced the number of military and civilian personnel. However, as Figure 1.1 shows, the share of money spent on goods and services for the Air Force has been trending upward, as have the shares of money spent on acquisition of new systems and large upgrades managed by system program offices (SPOs). The Air Force spends most of its dollars on weapon system acquisition and purchased goods and services, much of which comes from private-sector suppliers, making purchasing a critical capability for the service.[2] In fiscal year (FY) 2008, weapon system acquisition and purchased goods and services represented 71 percent of the total Air Force budget. While this aggregate percentage has varied between 40 and 71 percent over time, the percentage of the budget that is only for purchased goods and services has grown from a low of about 13 percent in 1950 to a high of about 44 percent in 2003.

The acquisition of weapon systems and purchased goods and services is integral to the Air Force's ability to execute its mission. As dollars become increasingly scarce and requirements seem to only grow, the Air Force needs to find ways to reduce the total costs of purchased goods and services while increasing performance.

In addition, within the past decade, the total Air Force expenditure for organic maintenance provided by government depots has been decreasing while the total expenditure for contract maintenance has been increasing, threatening to breach the 50:50 ceiling mandated

---

[1] *Purchased goods and services* includes all sustainment spending to support fielded weapon systems, base operating support, military housing and construction, services, activities outsourced to suppliers through A-76 competitions, transportation spending, goods acquired from the Defense Logistics Agency (DLA) or other non–Air Force organizations, and upgrades that are managed by system program managers (SPMs) at the ALC level.

[2] *Purchasing* includes all those activities required to obtain needed products and services from private-sector sources. These activities can be done directly by the Air Force or indirectly by other organizations on behalf of the Air Force, such as U.S. Transportation Command (USTRANSCOM).

Figure 1.1
The Air Force Budget, by Spending Category

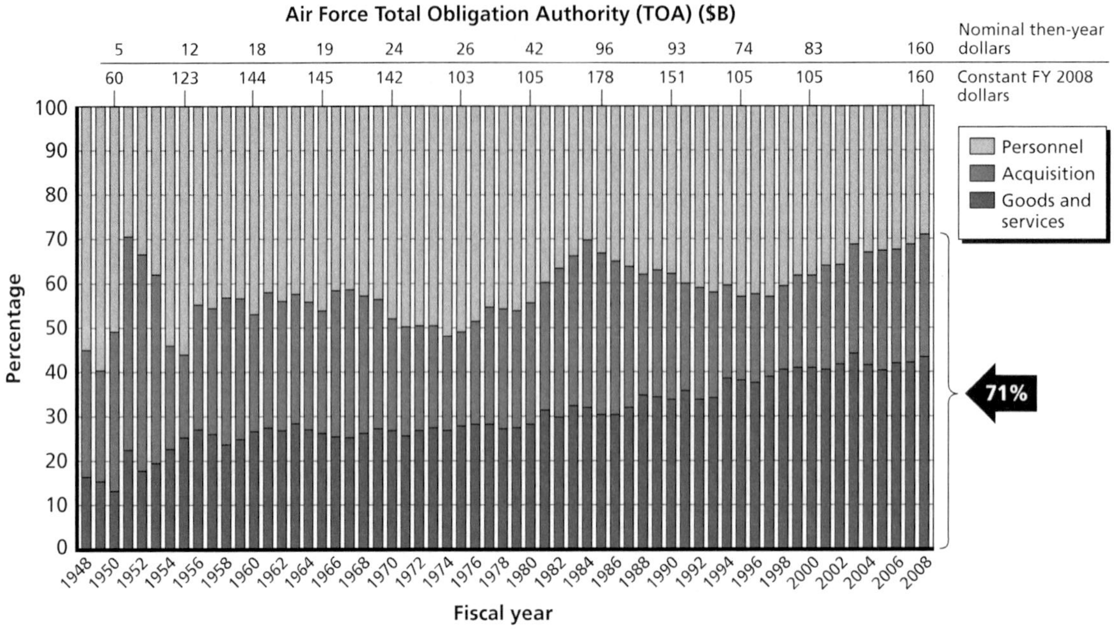

SOURCE: Office of the Undersecretary of Defense (Comptroller), 2009, Tables 6–18.
NOTES: The number of civilian personnel was extrapolated by multiplying U.S. Air Force (USAF) share of Department of Defense (DoD) civilian personnel (Table 7-5) by total DoD civilian personnel spend (Table 6-2).
RAND *TR904-1.1*

by law.[3] This is fueled both by growth in contract repair, such as contractor logistics support (CLS) and performance-based logistics (PBL) arrangements,[4] and by decreases in organic depot maintenance costs through retirements of older organically supported aircraft and depot transformation efforts to reduce costs and improve performance of organic depot-level activities.[5]

In response to the growth of purchases in the Air Force budget and the growth in contract repair within AFMC's depots, the Air Force implemented several initiatives intended to reduce the costs and increase the performance of its sustainment activities.

eLog21 is the Air Force's plan for transforming logistics business practices. A key part of eLog21 is Purchasing and Supply Chain Management (PSCM). PSCM is an Air Force pro-

[3] The "50:50" rule is found in Title 10, Subtitle A, Part IV, Chapter 146, Section 2466 of the U.S. Code and is titled "Limitations on the performance of depot level maintenance of material." It requires that each military service spend less than 50 percent of its total repair dollars on contract sources of repair, with the other 50 percent of dollars spent on organic maintenance. Going over this limit requires a waiver from the Secretary of Defense. The Air Force has gone above this congressionally mandated threshold in FY 2000 and FY 2001, when it spent more than 50 percent of its depot maintenance dollars on contract repair. (Discussions with Air Force Materiel Command [AFMC] Depot Maintenance Programs [AFMC/A4B], and AFMC Depot Maintenance Operations [AFMC/A4D] on October 30, 2006. See also Warren, 2000, and GAO, 2003.)

[4] Under CLS and PBL contracts, suppliers are rewarded for meeting customer outcomes, such as availability targets for parts, subsystems, or even entire weapon systems. Contractors also assume responsibility for most supply chain functions.

[5] Depot improvement efforts include implementation of best practices in Six Sigma and lean manufacturing to speed repair cycle time. Through lean manufacturing practices, shop layouts and processes have been reconfigured to minimize downtime and eliminate actions that add no value.

gram intended to improve and integrate AFMC's purchasing and supply processes. Its goals are to increase the availability of parts, improve product quality and delivery, and reduce the cost of doing business. An important component of PSCM is the Air Force's Supplier Relationship Management (SRM) program. SRM concentrates on identifying AFMC's most significant suppliers and building more effective, collaborative, and strategic relationships with them, with the aim of bringing material benefits to both supplier and customer.[6] Each of these transformational programs seeks to adapt proven private-sector practices for improving performance and reducing costs.

Although the Air Force began implementing SRM in 2002, implementation is moving more slowly than logistics leadership would like, and the cost improvements have been less than hoped for. The Air Force's concerns over the pace of performance and cost improvements with its key sustainment suppliers drove the request for this study.

Study Objectives and Research Questions

The goal of this study was to help the Air Force identify what it can do through its SRM initiatives to improve contract depot-level performance in spares, repairs, and other services and drive down total costs. The study had two objectives: (1) identify best practices in supplier relationship management and development that have reduced costs and improved performance in other organizations; (2) highlight any impediments to applying these best practices within the Air Force and recommend ways to adapt them to the Air Force environment. This study addressed four research questions:

1. What SRM practices has the Air Force implemented to reduce its total supply chain costs and improve performance?
2. What are the best practices in SRM in use by the private sector?
3. What, if any, factors potentially limit the implementation of best practices in SRM within the Air Force? If impediments do exist, what must change for these practices to work effectively, and how can they be addressed?[7]
4. How can AFMC and the Air Force know whether SRM is working? With so many other initiatives underway, how can the benefit of SRM be detected and measured?

---

[6] Other components of PSCM include working more effectively as an organization to provide parts through commodity councils (CCs) and customer relations management. CCs develop sourcing strategies for groups of National Stock Numbers (NSNs) with like characteristics, such as engine parts, and write and manage repair contracts. The purchasing of spares and depot-level contract repairs have been centralized to eight CCs located at the three ALCs. The eight CCs are Aircraft Accessories, Aircraft Engines, Aircraft Structural, Communications-Electronics, Instruments, Landing Gear, Secondary Power Systems, and Support Equipment. Responsibility for writing and managing spare contracts moved from the military services to DLA in the 2005 BRAC decision. Customer relations management provides a single interface between AFMC and the customer, which gives the customer a more efficient means of communicating with the wholesale system.

[7] Implementation of best supplier development practices could encompass policies, processes, data systems, training, workforce skills, and organizational, cultural, and legal issues.

Methodology

Our research methodology had three components. First, we reviewed the academic, business, and defense literature.

Second, we conducted interviews with knowledgeable government personnel and managers at commercial firms. We interviewed government personnel at AFMC and other locations about current AFMC SRM practices. We had discussions, either face-to-face or by telephone, with the five Strategic Supply Relationship managers (SSRMs) who lead the SRM teams, as well as personnel from the Defense Contract Management Agency (DCMA) and DLA.[8] A summary of those interviews is shown in Table 1.1.

To target our commercial firm interviews, we used our literature review to identify private-sector exemplars in supplier cost management. We identified commercial examples where supplier development has been successful and, to the extent that the procedures are reported, how. We selected two companies, basing our choices on the companies' reputation for SRM and on their willingness to be interviewed. One company is involved in automotive manufacturing,

Table 1.1
Supplier Relationship Management Interviews

Organization	Type of Individual(s)	Number of Interviews	Total Number of Personnel Interviewed	Number of Individual Civilian SRM Team Members Interviewed
Government				
AFMC/PK (Contracting)	SSRM	2	2	1
AFMC/A4 (Logistics)	SSRM	1	2	
Oklahoma City Air Logistics Center (OC-ALC)	SSRM, SRM team	1	5	2
Ogden Air Logistics Center (OO-ALC)	SSRM, SRM team	3	7	4
Warner Robins Air Logistics Center (WR-ALC)	SSRM, SRM team	1	4	2
DLA/J-3 (Operations)	Strategic supplier alliance	1	3	
DCMA/OCX (Performance Integration)		1	1	
Commercial				
Companies	SRM leaders and senior managers	2	2	
Total		12	26	9

NOTES: Individual SRM team member civilian interviews, not including SSRMs, were with GS-12 and GS-13 personnel. GS = General Schedule.

8 We interviewed the three SRM teams and their SSRMs at ALCs as separate groups and both SSRMs at AFMC headquarters as individuals. On other occasions, we individually interviewed 11 SRM team members: nine civilians and two contractors. Finally, we conducted a group interview with DLA and an individual interview with DCMA. For more on our AFMC SRM interviews, see Appendix C.

and the other is engaged in aerospace manufacturing. Both have been publicly recognized for the quality of their relationships with suppliers. Using the information from our AFMC interviews and the literature review, we developed a protocol for interviewing personnel at these companies who are knowledgeable about their company's supplier development activities. In our interviews, we sought to understand the kinds of analyses that preceded commercial supplier development initiatives, how they were accomplished, the essential elements of success, and skills required for supplier development teams. We also interviewed selected suppliers that have received supplier development assistance.[9]

Third, we analyzed Air Force item-level detailed contract transaction data for FY 2006–2008 from DD350 and Federal Procurement Data System–Next Generation (FPDS-NG) data systems for SRM supplier spending.[10] We gathered detailed contract data on the top 21 SRM suppliers based on NSN spending for spares and repair.

Finally, using the information from our reviews, interviews, and analyses, we identified impediments to implementing best SRM practices in the Air Force and developed recommendations on how to address them.

Several important developments have occurred since this research was conducted. AFMC has centralized the management of its supply chain (e.g., spares and repair assets) with the creation of the Air Force Global Logistics Support Center (AFGLSC). Consequently, the CCs, SRM teams, and contracting officers have been reorganized. AFMC has also reorganized from wings/groups/squadrons to directorates/divisions/branches. Organizational names used here are those that existed at the time of this research. Also, DLA is now responsible for writing contracts for new reparables instead of the military services. We note these changes where appropriate.

How the Report Is Organized

The remainder of this document is organized as follows. Chapter Two describes current best practices in SRM. Chapter Three explains our findings on current SRM practices at AFMC. Chapter Four describes those challenges and hindrances that slow or frustrate the Air Force's implementation of SRM. Chapter Five presents our conclusions and recommendations for improving the implementation of SRM in the Air Force.

---

[9] In each case, we agreed not to disclose the names of the individuals interviewed or the names of their companies.

[10] DD350 refers to data from Individual Contracting Action Report forms; this form is also known as DD350.

CHAPTER TWO

Current Best Practices in Supplier Relationship Management

This chapter contains three parts. First, it shows the benefits that accrue when an organization develops its suppliers. Second, it discusses how organizations can execute SRM well, i.e., the best practices. Third, it describes a model for developing suppliers.

Best PSCM practices, of which SRM is a key component, pertain to the entire supply chain and involve the following sequential practices:

1. conducting spending, market, and risk analyses and developing supply strategies
2. rationalizing the supply base and consolidating contracts
3. establishing long-term relationships with best suppliers
4. helping key suppliers improve quality, cost, and performance
5. integrating key suppliers into the organization.

The first three steps are often referred to as *strategic sourcing*.[1] SRM focuses on the last two practices.

Why Implement SRM? The Benefits of Developing Suppliers

A number of publications have discussed the initial and ongoing benefits of improving PSCM practices (e.g., Moore et al., 2002). The first three steps listed above (i.e., strategic sourcing) typically produce one-time savings, while the latter two steps (i.e., SRM) can, if done well, lead to continuous or year-after-year savings. SRM does this by continually working with suppliers to identify and take costs out of the end-to-end supply chain and improve quality and other aspects of performance. As a representative of one company we interviewed said, SRM is about reducing the total cost of doing business with each of your key suppliers.[2]

Figure 2.1, which is adapted from Nelson, Moody, and Stegner (2001) and based on actual cost data, shows the potential long-term savings from SRM. The Producer Price Index (PPI) line shows the cumulative percentage growth in costs of a group of automotive goods over the 1992–1998 time period. The "good company" line shows the cost growth experienced

---

1 "Strategic sourcing is the *collaborative* and *structured* process of critically analyzing an organization's spending and using this information to make business decisions about acquiring commodities and services more effectively and efficiently" (Johnson, 2005).

2 Boeing identifies its "key suppliers" based on annual spending and unique business requirements. These suppliers follow a more frequent evaluation schedule (semiannual or more frequently if a program or site feels it is necessary) and are rated by all programs/sites conducting business with them in the months of April and October. (See Boeing, no date.)

Figure 2.1
SRM Can Lead to Large Long-Term Savings

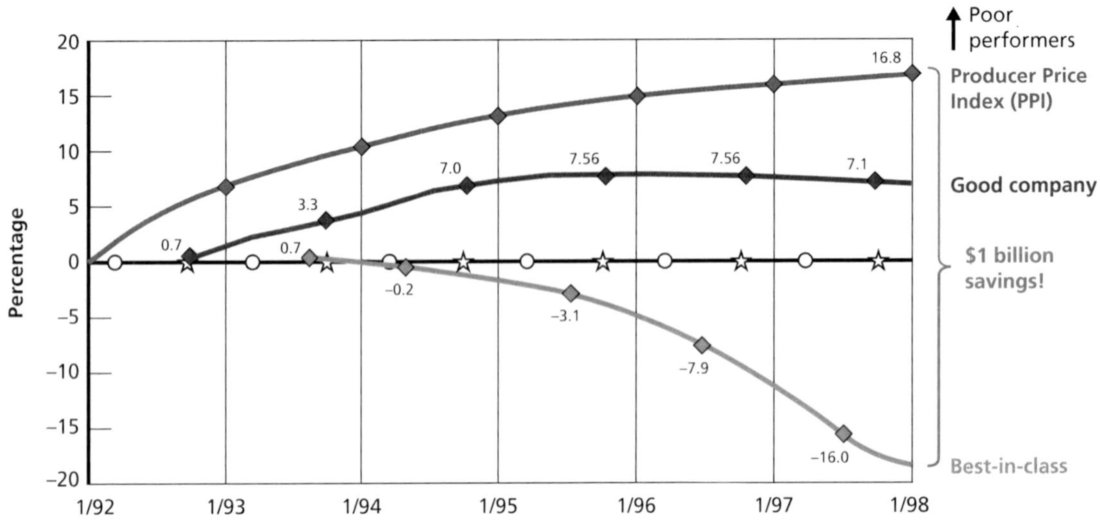

SOURCE: Nelson, Moody, and Stegner, 2001. Used with permission.
RAND *TR904-2.1*

by a good U.S. automotive company that had adopted some best PSCM practices but did not focus on working with suppliers to reduce costs and hence prices. The "best-in-class" line shows cost trends experienced by Japanese transplant companies[3] during this time period. These best-in-class firms paid much less than the industry average PPI compared with the good U.S. company, because they paid more attention to fostering close relationships with their suppliers. This difference between the good U.S. company and the best-in-class transplant companies translates to roughly $600 million more in savings during this time period.

However, SRM involves much more than reducing prices relative to the PPI. It is about lowering the *total cost of ownership* (TCO) of the end-to-end value chain and increasing value to the buying enterprise and, ultimately, the end customer.[4] Some of the non-price costs associated with a buying transaction that can occur before, during, or after a transaction can be significant, in some cases matching or even exceeding the nominal price.

Examples of non-price costs that SRM should consider include the following:[5]

- internal business cost
- transportation
- warehousing

---

[3] Japanese transplant companies are major business units of companies with headquarters in Japan, such as Honda, Nissan, and Toyota, that manufacture products in the United States.

[4] Almost 67 percent of 15 business units of 14 aerospace/defense enterprises reported that their supply management organization tracks total cost savings. See CAPS Research, 2007. According to Feller, Shunk, and Callerman (2006), "the primary difference between a supply chain and a value chain is a fundamental shift in focus from the supply base to the customer. Supply chains focus upstream on integrating supplier and producer processes, improving efficiency and reducing waste, while value chains focus downstream, on creating value in the eyes of the customer." See Feller, Shunk, and Callerman, 2006, p. 4.

[5] See Chapman et al., 1998; Ellram, 2002; and Bowersox, Closs, and Cooper, 2002.

- inventory carrying cost
- purchasing administration
- factory yield
- damaged field product
- joint supplier/company/customer life cycle cost
- production capacity
- research and development (R&D)
- specifications
- expediting.

Companies such as Honda, John Deere, and Praxair are increasingly making full supplier cost disclosure a condition for consideration as a potential supplier while also corroborating data from multiple sources, particularly unbiased external data and their own knowledge of internal costs. This helps in identifying TCO and targeting SRM initiatives (Ellram, 2002).

SRM is a major source of many PSCM performance improvements reported in the academic, business, and trade literatures, particularly those improvements that are continuous in nature. The improvements tend to be wide-ranging and include such areas as more-responsive deliveries, reduced inventory requirements coupled with higher service levels, higher quality, and more-rapid product development.

A survey of leading companies found a 7–10 percent *per year* average improvement in delivery responsiveness attributable to supplier development and management initiatives (Trent and Monczka, 1998). For example, Sun Microsystems' Enterprise Services organization reported improvements in repair-parts-vendor turnaround times from 34–40 days to 4–5 days after it fully integrated its technology platform and systems with those of 12 of its key vendors in the Americas (Pazmany, 2000). Rockwell Collins reported that supplier on-time delivery increased from 83.8 percent to 96.5 percent in three years after it developed a portal through which suppliers can gain access to engineering designs and forecasts, download scorecards that monitor their performance, and send invoices to and receive payment from Rockwell Collins.[6]

Companies also report requiring less inventory and obtaining higher service levels as a result of SRM. For example, Cessna reported a 113 percent increase in production inventory turns over six years, with dramatically higher material availability. Cessna achieved this improvement by creating a strategic plan and cross-functional commodity teams that rationalized the company's supplier base; developing a Maturity Path Development tool that aligns supplier strategy with that of Cessna's; revamping its sales, inventory, and operations plan to improve performance to customer expectations and reduce inventory turns; implementing use of Malcolm Baldrige National Quality Award criteria and Six Sigma quality tools to drive improvement in supplier performance; and introducing a value analysis/value engineering process that encourages supplier involvement in removing cost from the supply chain (Avery, 2003).

Companies also reported 10–13 percent per year average improvements in quality due to supplier-related initiatives (Trent and Monczka, 1998). For example, Honda of America Manufacturing Inc. (HAM) reported a reduction in parts defects from 7,000 parts per million (ppm) to less than 150 ppm in 12 years (Fitzgerald, 1995; Nelson, Mayo, and Moody, 1998). In

---

[6] In addition, cross-functional commodity teams meet frequently with top suppliers to share technology roadmaps, and suppliers serve on a Supplier Alliance Advisory Council, providing valuable feedback to material and supply (Avery, 2005).

1990, as a part of its BP[7] program, a team of four HAM associates would work at a supplier for three months, focusing on improving one production line. HAM also held its first BP conference with suppliers that year, in which some suppliers presented their BP activity to the rest of the supply base. After working with suppliers on their production processes, HAM eventually moved its BP program upstream to suppliers' product development process. The BP program was so successful that HAM developed an offshoot program—Best Quality—that focused on improving the quality of key safety-related or highly visible parts and on suppliers with the best opportunity for quality improvement. By 1995, HAM had about 800 associates[8] working full-time with its 320 suppliers. HAM's purchasing unit constantly rates and compares suppliers and gives annual awards to its top-performing suppliers.

Lastly, companies reported an average 22 percent reduction in product development time over eight years due to supplier-related initiatives (Trent and Monczka, 1998). For example, Alcatel reports that it reduced its new product development time by about six months over five years and saved millions in cost by involving buyers and suppliers in design (Carbone, 2004c). General Electric (GE) reported that its jet engine development time went from 60 months to 28 months and that it dramatically reduced the cost of developing new engines. GE changed where and how it builds engines and brought in several new partners, including competitors, to share development costs, allowing GE to take on more projects (Siekman, 2002).

Toyota's development of Garden State Training (GST) provides a specific example of the kinds of improvements that can be achieved from supplier development. Toyota sent a team of production system specialists to GST to work on moving equipment on the shop floor into one-piece (i.e., single-piece) flow cells,[9] implementing *kanban* systems,[10] developing standardized procedures for cutting leather, and slashing inventories of cowhide by more than 64 percent. Over two years, emergency air shipments fell to nearly zero, quality defects were almost eliminated, productivity more than doubled, and lost-time accidents declined by 99 percent without GE expending any capital (Licker and Wu, 2000).

Not all benefits of supplier development efforts are as easy to identify and measure as these examples indicate, and benefits can vary in their scope and degree of success. Krause and Handfield (1999) grouped reported improvements from supplier development efforts into the following three major categories:

- suppliers' performance and capabilities
  - order cycle time
  - quality
  - delivery

[7] "BP stands for Best Position, Best Productivity, Best Product, Best Price, and Best Partners. It is a powerful continuous improvement process designed by Honda." See Nelson, Mayo, and Moody, 1998, p. 4.

[8] About 300 are in purchasing, 200 are in quality, and 300 are manufacturing and production engineers. See Fitzgerald, 1995.

[9] These are work areas where machines of different types performing different operations are arranged into a tight sequence, typically a U-shape, to permit single-piece flow (i.e., products flow, one at a time, through various operations without interruptions, backflow, or scrap) and flexible deployment of personnel who work on multiple machines. See Womack and Jones, 1996.

[10] A *kanban* system uses small cards attached to boxes of parts to regulate downstream pull (i.e., just-in-time) production in the Toyota Production System by signaling that upstream production and delivery has occurred. See Womack and Jones, 1996.

  - price change
  - reduced total costs
  - new product development time
  - R&D capabilities
- relationships between the buying enterprise and its suppliers
  - increased levels of trust
    - improved access to technology
  - greater levels of integration
    - technology and growth plans
    - information technology systems
  - improved supplier access to help from buying enterprise, if needed
- buying enterprise's competency in managing suppliers.

Table 2.1 summarizes results from a survey of 31 U.S. and foreign companies that primarily focused on the first category of SRM efforts, i.e., suppliers' performance and capabilities (Krause and Handfield, 1999). Note that the greatest reported improvements are in order cycle-time (19 percent), quality (24 percent), on-time delivery (39 percent), and reductions in new product development time (19 percent), while the least significant improvements were price changes (3 percent) and shared cost savings (7 percent).

Best Commercial Practices for Developing Suppliers

Our analysis of commercial practices led us to identify what we regard as the best SRM practices. We identified the following six practices:

- Manage total business with each supplier.
- Measure and shape supplier performance.
- Involve key suppliers early in product design.
- Host high-level meetings that promote dialogue with suppliers.
- Recruit skilled personnel.
- Develop personnel so they have a thorough knowledge of suppliers.

Table 2.1
Reported Improvements from Supplier Development Efforts

Area of Improvement	Percentage Improvement
Order cycle time (order placement to receipt)	19
Quality (fewer defects, warranty returns)	24
On-time delivery (within buyer's specified window)	39
Percentage of supplier's price change	3
Shared price reduction	7
New product development time (concept to volume production)	19
Access to new technologies	15

SOURCE: Krause and Handfield, 1999.

We describe each of these practices in more detail below.

Manage Total Business with Each Supplier

When most large enterprises analyzed their spending, they learned that they had many contracts with some of their suppliers, which reduced their leverage and made it more difficult to measure and manage supplier performance or work with suppliers to improve performance and reduce total costs.[11] Indeed, companies recognized for their excellence in PSCM manage their total business with each supplier. They do this by consolidating their contracts and relationships with each supplier or supplier's business units and locations into one or very few contracts, and they link future business to supplier performance on all contracts and relationships with the supplier. They claim that suppliers typically prefer the large-volume, more stable production business to the lower-volume, more variable and uncertain service parts business. As an example, these companies retain competition among their suppliers until selection for production. After production has begun, service parts are often sole-sourced to the original manufacturer. Consequently, at the same time they negotiate production, some companies will negotiate for service parts support to continue beyond production, for the expected useful life of their product.[12] In addition, companies use supplier past performance on service parts as well as production contracts as part of the selection criteria for awarding future production business.[13]

These companies also monitor and shape the capabilities and capacities of their "key" suppliers. They do this by guaranteeing business as long as suppliers continue to improve and meet expectations, by sharing technology and other plans so suppliers can better plan their capacity and technologies, and by working with suppliers to develop new capabilities, such as supplying production facilities in other regions, doing their own R&D, or expanding to similar or related products.

Enterprises recognized for their best PSCM practices also try to balance competition, supplier dependence, and risks. They want to form close relationships with their key suppliers, but they also want to make sure those suppliers continue to improve. These companies carefully manage long-term competition so that it provides incentives for the supplier to improve continuously while not thwarting the building of trust and other benefits that develop in longer-term relationships, such as communication, cooperation, integration, and joint improvement initiatives. Thus, for example, while Toyota uses one supplier per part for each model, it typically has more than one supplier across models, varying total business with each supplier according to their performance. When a rare crisis occurs, Toyota works hand in hand with suppliers to get through the crisis (Licker, 2004). For example, when a piston ring maker, Riken Corporation, closed 11 parts factories after an earthquake in northwestern Japan, Toyota dispatched more than 200 engineers to help Riken resume production (Ohnsman, 2007).

Some leading manufacturers that have separate production and service parts support divisions, with separate SRM personnel for their products, require that SRM personnel in their production and service divisions work toward common goals, so that they are speaking to suppliers as "one voice" and are not working at cross-purposes. However, depending on the specific supplier issue, one or the other division will lead the supplier improvement effort. For example,

---

[11] For an excellent example of this, see "Gloves on the Boardroom Table," by Jon Stegner, in Kotter and Cohen, 2002, p. 29.

[12] This "life of part" contracting is a part of Cessna's new product development process. (See Morgan, 2000.)

[13] Boeing tracks supplier performance data monthly by each Boeing site. (See Boeing, no date.)

one company we interviewed told us that, depending on the specific supplier issue (e.g., quality versus packaging), either production or service parts SRM personnel will work with the supplier to resolve the issue.

Measure and Shape Supplier Performance

In addition to managing total business with suppliers, companies recognized for their best PSCM practices regularly measure and shape their suppliers' performance and the supply base. They do this in several ways. First, they establish a supplier performance measurement system that regularly—often electronically—measures supplier performance along dimensions key to their strategic goals. For example, Cessna's Supplier Tracking and Rating System (STARS) measures the quality/reliability (i.e., defect rate in ppm and field performance), schedule (receipts on time), and cost (annual productivity improvements, market competitive, fixed pricing, or escalatory) of its suppliers (Cessna, 2007). Boeing's Enterprise Supplier Tool (BEST) measures its suppliers on quality, delivery, and general performance assessment (in areas of management, schedule, technical, cost, and quality) (Boeing, no date).

Next, buying enterprises rank their current suppliers into three to five overall performance groups using a formula for weighting each supplier's score for each performance and cost metric.[14] Once suppliers have been segmented into overall performance groups, buying enterprises set future performance targets, which are periodically raised for each group to keep suppliers improving in all key areas.

Lastly, many buying enterprises reward suppliers for superior performance, such as exceeding performance targets, with awards and/or additional business. As examples, Cessna's STARS Supplier Excellence Award and Boeing's Performance Excellence Award[15] are presented by the buying enterprise's top management at annual supplier awards dinners.

Suppliers with top ratings/performance are often designated as preferred suppliers and rewarded with additional business and less oversight. For example, suppliers that receive the Boeing Performance Excellence Award not only get a trophy suitable for a lobby display, they also get source selection preference based on performance levels (Boeing, no date).

Conversely, if supplier performance falls below certain standards, the supplier is often required to develop a "get well" plan,[16] which is regularly monitored. Companies will sometimes send teams to help out suppliers. The capability to help suppliers in trouble is particularly important for companies that have moved to lean, just-in-time operations with little inventory, because they need to help suppliers solve any problems very quickly to prevent or minimize adverse effects on production. For example, if a supplier puts a Toyota assembly plant in danger of shutting down, Toyota will send a team of people to analyze the supplier's plant operations,

[14] For example, Cessna's five supplier groups are outstanding, superior, acceptable, marginal, and unacceptable, while Boeing's five supplier groups are gold, silver, bronze, yellow, and red. Altel has four categories of suppliers: strategic, preferred, allowed, and restricted (Carbone, 2007). Toyota rates suppliers from one (e.g., a plant burns down) to five (exemplary Toyota Production System supplier) (Licker, 2004).

[15] This replaced the Boeing Preferred Supplier Certification program as of September 30, 2007. As stated in Boeing (2007), the eligibility criteria are

- composite performance rating at Gold or Silver level for 12 consecutive months from October 1 through September 30
- annual contract payment of $100,000 or greater
- minimum of one receipt for 10 of the 12 performance months.

[16] A supplier action plan identifies the root cause of a performance or cost problem and quickly fixes it so that it does not recur.

and the supplier must develop an action plan to address all of their concerns (Licker, 2004).[17] For example, when one if its suppliers, Northwest Tool & Die, faced bankruptcy, HAM came to the rescue, teaching Northwest Tool & Die a more efficient way to design and manufacture and steering millions of dollars of new business to the supplier. HAM's intervention helped Northwest Tool & Die exit bankruptcy and reduce its prices to within 15 percent of its low-cost Asian competitors (Muller, 2007).[18] And, as mentioned earlier, when an earthquake hit Japanese piston ring maker Riken Corporation and halted production, Toyota dispatched more than 200 engineers to help it resume production (Ohnsman, 2007). Suppliers with low or poor ratings/performance that do not improve are usually phased out if viable alternatives exist, or helped to improve if they do not.[19]

Firms such as Toyota and Cessna continually challenge their suppliers to improve. Their supplier development programs include aggressive targets for performance and cost improvements and challenges to meet those targets (Licker, 2004; Cessna, 2007).

Involve Key Suppliers Early in Product Design

Many companies known for their SRM involve key suppliers early in their product design—a practice called early supplier involvement. For some industries, as much as 90 percent of the cost of new products and much of the manufacturing quality challenge is committed at the design stage when decisions about materials, specifications, manufacturing processing, and packaging are made (Nelson, Moody, and Stegner, 2005, p. 104). The design stage is also the best time to balance needs for innovation and reductions in total costs. Firms practicing SRM want to leverage their suppliers' design capabilities, knowledge of manufacturability, and innovation to improve the product and TCO. For example, Honda will "invite" some suppliers to locate guest designers within Honda's facilities and work side by side with Honda's resident engineers, designers, and technologists in the early stages of a new project (Laseter, 1998). It is at the design stage that companies try to reduce complexity and costs through parts standardization and reuse across products as well as improve manufacturability. For Boeing's new 787 *Dreamliner* aircraft, major suppliers will design and build entire sections of the plane, shipping them to Boeing for final assembly and testing at Everett, Washington (Bowman, 2007).[20]

Host High-Level Meetings That Promote Dialogue with Key Suppliers

Best SRM practices include hosting annual, high-level customer/supplier meetings. These meetings are often called supplier conferences or supplier councils, and their purpose is to demonstrate mutual commitment to the relationship, promote dialogue on expectations and

---

17 Toyota operates with minimal inventory, so parts problems are quickly identified and quickly fixed.

18 Honda has a philosophy of locating production plants in the markets where there is a demand for its products and using local resources where they are competitive. It encourages suppliers that understand the "Honda way" and are willing to improve their competitiveness. See Honda, no date. Honda seeks long-term, mutually beneficial supplier relationships and will commit the necessary effort and resources to help suppliers reach a desired performance level. See Fitzgerald, 1995.

19 The loss of future business may be the strongest incentive available to a buyer for a supplier that cannot be phased out and shows an unwillingness to improve.

20 Suppliers provide up to 70 percent of the Boeing *Dreamliner*'s content as subsystems, not parts. They pay their own upfront costs related to engineering, facilities, equipment, and tooling; design parts of the plane as well as build parts; integrate subsystems from their own second-tier suppliers; test and verify all systems; and collaborate and meet regularly to discuss engineering and other issues and plan strategies to resolve them (Teague, 2007). Aircraft production has been delayed by supply chain problems that are being addressed (Lundsford and Michaels, 2008).

ways to improve, share future plans and technology roadmaps, plan and highlight improvement efforts, and present supplier awards. For example, suppliers meet each year at one of Honda's plant sites for its Annual Supplier Conference. They hear presentations from the president of the company about Honda's upcoming strategy and review quality and delivery performance against goals. Harley-Davidson and John Deere have similar annual supplier conferences (Nelson, Moody, and Stegner, 2001). One company also told us that these conferences conveyed who its preferred suppliers were, as attendees could observe which ones were no longer invited.

Recruit Skilled Personnel

Leading companies recruit high-quality personnel with an array of skills for their SRM positions.[21] SRM personnel are often recruited from engineering, manufacturing, and purchasing functions. In addition to these specialized skills, firms seek out people with good "soft" skills, such as communications and problem solving. Other valued expertise includes knowledge of the supply chain, process-improvement techniques (e.g., lean, Six Sigma, Total Quality Management [TQM]), and the enterprise's internal ordering and receiving processes, communication interfaces and data systems. This knowledge of the buying enterprise's internal processes and practices helps SRM personnel interface and communicate with suppliers.[22] Buying enterprises feel they can train personnel on the technical aspects of SRM, such as negotiating skills. At several companies we interviewed, we were told that positions on supplier development teams were highly sought after because personnel enjoy setting their own priorities, visiting suppliers, and helping suppliers fix problems and improve.

Nelson, Moody, and Stegner (2005) claim that investments in supplier development and supply base management yield huge payoffs. They report that at Deere and Delphi, a $100,000 investment in one supplier development engineer often yields a return of three to ten times that amount and that top performers can achieve even more. As discussed earlier, these benefits, which are hard to quantify, come from improvements at suppliers or in supply chain processes that reduce total costs or improve quality or delivery.

Develop Personnel So They Have a Thorough Knowledge of Suppliers

SRM personnel need to know suppliers' processes, costs, capacities, and capabilities. They need to be able to work with and help suppliers develop or fix processes to meet current needs as well as continually improve to meet future needs.

---

[21] In a 2006 survey, almost 77 percent of 15 business units of 14 aerospace/defense enterprises reported that the supplier development relationship was included in their supply management organization. Over 45 percent reported that supplier quality was included in their supply management organization, and almost 64 percent reported that supplier management was. Seven business units reported that a little over 7 percent of supply management personnel were assigned to supplier development, which is an increase from the 3 percent reported in a 2003 survey. Four business units reported that almost 27 percent of supply management personnel were assigned to SRM (CAPS Research, 2007). In a 2003 survey of 17 business units of 10 aerospace/defense enterprises, over 94 percent reported that management/control of supplier development/relationships was in the supply management organization and that the average number of full-time personnel performing these activities was 27 (CAPS Research, 2005).

[22] One company we interviewed stressed active communication and listening skills as well as an aptitude for being analytical and customer-focused. This company also said that it seeks people who are professional and polite.

A Model for Developing Suppliers

Here, we describe a model for developing world-class suppliers and a world-class supply base that resembles the process being followed, as a whole, by a group of companies striving to build a globally aligned network of suppliers (Krause and Handfield, 1999). We also review factors that facilitate or inhibit SRM.

Most of the research on supplier management and development involves case studies of specific situations. Krause and Handfield (1999) set out to describe a process that enterprises employ to develop a world-class supply base. They reviewed the literature on and visited organizations involved in developing a world-class supply base, surveyed buying firm representatives who were members of the National Association of Purchasing Management (NAPM), and documented a number of best practices, which they compared, refined, and integrated into a model of the core processes required to develop a world-class supply base.[23] They observed that developing a world-class supply base occurs in four major stages, each of which requires increasing levels of commitment and resources and delivers a higher level of benefits to the buying enterprise and its suppliers over time. Figure 2.2 displays the Krause and Handfield Supplier Development Model.

The first stage—identify, assess, and rationalize the supply base—occurs before supplier development. The other three stages of the model are classified as supplier development. With each stage of the model, the benefits increase and suppliers are increasingly integrated with the buying enterprise (Krause and Handfield, 1999).

We briefly describe Krause and Handfield's four stages and five critical success factors here and refer the reader to a more detailed discussion of these stages in Appendix C.

Stage 1: Identify, Assess, and Rationalize the Supply Base

The goal of Stage 1 is to rationalize the supply base to those suppliers who are better aligned with the buyer's strategic needs and have the wherewithal and commitment to continually improve and invest in the relationship. This stage is focused on the buyer identifying its strategic supply chain needs; searching for the best suppliers worldwide; developing measures for monitoring supplier quality, cost, and performance; and, after assessing suppliers, deciding which to focus more of the business on and develop, which to replace, and which ones to phase out.

---

[23] The case study firms that Krause and Handfield visited were in the automotive and electrical/electronics industries. Krause and Handfield interviewed senior purchasing executives (i.e., manager [division], vice president of purchasing, quality manager, or in some cases, a dedicated supplier development manager). Additional interviews, as appropriate, were done with engineering, operations, quality, and logistics personnel. Their in-person interviews included four companies in each of the two industries in the United States and in the United Kingdom, four companies in the automotive and three in the electrical/electronics industries in Japan, and three companies in each industry in South Korea (Krause and Handfield, 1999).

Krause and Handfield's Survey respondents included 173 in the automotive industry, 70 in the electrical/electronics equipment industry, 59 in the machinery industry, and 61 miscellaneous industries, for a total of 363 respondents. Most of the respondents were at companies with gross sales greater than $10 million, with 13 percent at companies with annual gross sales over $1 billion (Krause and Handfield, 1999).

NAPM became the Institute for Supply Management (ISM) in January 2002.

Figure 2.2
Krause and Handfield Supplier Development Model

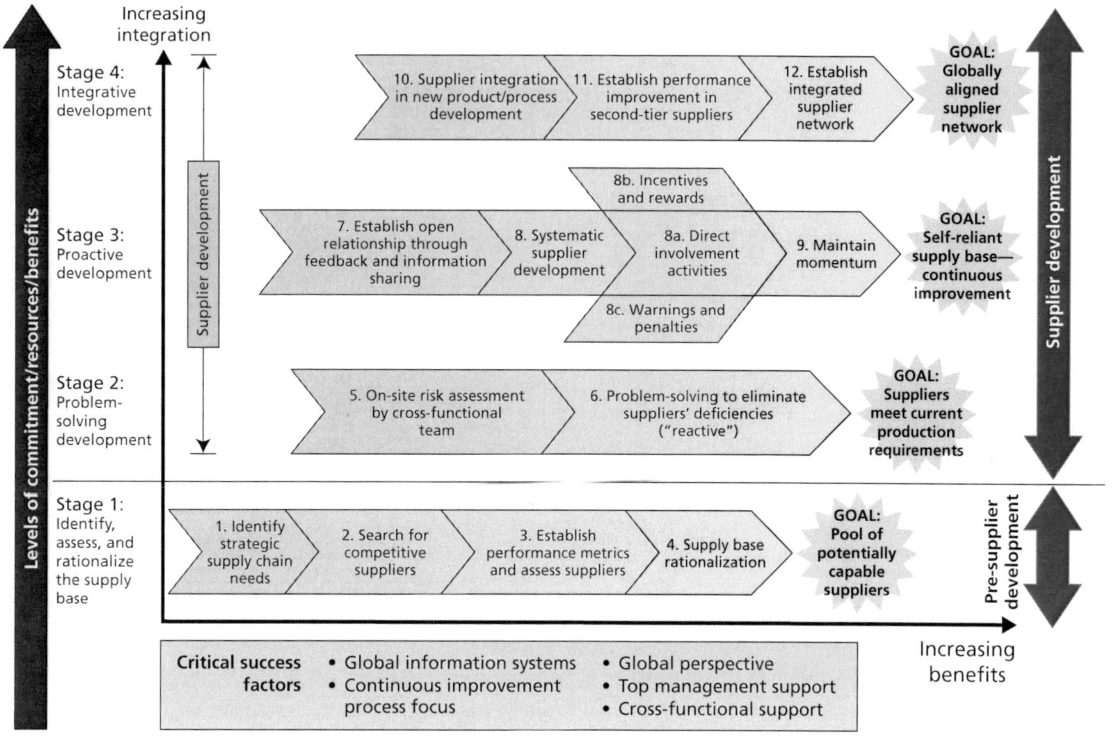

SOURCE: Adapted from Krause and Handfield, 1999. Reprinted with permission from the publisher, Institute for Supply Management and W. P. Carey School of Business at Arizona State University.
RAND *TR904-2.2*

Stage 2: Problem-Solving Development

The goal of Stage 2 is to work with suppliers to meet the buying enterprise's current production requirements. The buyer develops a thorough understanding of the supplier's capabilities, strengths, and weaknesses/deficiencies as well as its ability to meet future requirements. The buyer might do this through on-site visits with cross-functional teams using a risk assessment measurement protocol to identify any problems with the supplier's processes, capacities, or capabilities that affect the buying enterprise's production and require immediate attention.

Stage 3: Proactive Development

The goal of Stage 3 is a self-reliant supply base that continually improves to meet evolving buyer needs. The buyer focuses on putting in place the incentives and mechanisms for suppliers to continually improve and invest in those areas deemed most likely to benefit performance and relationship to avoid problems before they occur. The buyer does this by establishing open relationships, through feedback and information-sharing and through systematic efforts to develop the supplier. These efforts can be direct involvement activities, such as *kaizen* events,[24]

[24] Womack and Jones (1996) define *kaizen* as "Continuous, incremental improvement of an activity to create more value with less *muda*. Also called point kaizen and process kaizen." They define *muda* as "any activity that consumes resources but creates no value," i.e., waste.

incentives and awards, and warnings and penalties. Undergirding all of these efforts is also the need to maintain momentum to prevent any backsliding or losing ground.

Stage 4: Integrative Development

The goal of Stage 4, the last stage of Krause and Handfield's model, is a globally aligned supplier network. Though few buying enterprises have achieved this level, it has the potential to deliver the greatest benefits. In this stage, the first-tier suppliers are integrated in new product and process development and the buying organization begins efforts to extend supplier performance improvement to lower tiers in the supply base. The buyer also wants suppliers to align their own strategic plans for growth in the areas and locations where it plans to operate and compete.

Critical Success Factors

Krause and Handfield also identify five critical success factors that are necessary for the successful implementation of SRM:

- **Top management support.** The support of the buying enterprise's and suppliers' top management is important to gain resources, schedules, and formal agreements for unilateral and joint supplier development efforts.
- **Cross-functional support.** Involvement by a number of different functional personnel from the buying enterprise is needed to perform supplier risk and other assessments, set up solutions, and monitor progress. These functions can include purchasing, quality assurance, engineering, materials management, and manufacturing, among others.
- **Continuous improvement process focus.** A culture of continuous improvement, initially at the buying enterprise and eventually at key supplier enterprises, is critical to the success of the supplier development model and SRM in general. A number of techniques have emerged from the business literature to help enterprises design quality into their products and processes, control their processes, and improve them. These techniques can provide a common language for communicating within enterprises across locations, functions, and management levels as well as among supply chain partners on the need for change and ways to effectively obtain improvements.
- **Global perspective.** Competition and the global, interconnected economy are driving enterprises to think and act more globally. Some buying enterprises are requiring suppliers to supply the enterprise's global facilities equally well and are measuring suppliers' performance globally. A global perspective widens the pool of prospective suppliers and increases incentives for suppliers to continuously improve to gain and maintain the higher volume business.
- **Global information systems.** As enterprises and their supply bases become more globally dispersed, global information systems are required to effectively communicate within and among them. Such systems can communicate forecasts, production requirements, design changes, and forthcoming solicitations. They can also serve as a repository of enterprise knowledge on supply markets (e.g., capacities, prices, and trends); suppliers (e.g., availability, quality, and design capabilities); local laws, regulations, and customs; and competitors.

The next chapter turns to the Air Force's implementation of SRM within AFMC.

CHAPTER THREE

Current Supplier Relationship Management Practices at Air Force Materiel Command

We now turn to how AFMC has implemented SRM and the best practices it uses. We begin by describing the key players in the SRM process, what they do, and how they relate to one another. We then discuss the implementation of SRM in AFMC and AFMC's specific practices.

Since this study was conducted, the organizational structure that existed when SRM was first implemented has changed with (1) the creation of the AFGLSC, which centrally manages spares and reparables and oversees the CCs and SRM teams; (2) the implementation of the 2005 BRAC, which reassigned responsibility for writing and managing spares contracts from the military services to DLA; and (3) the reorganization of AFMC from wings/groups/squadrons to directorates/divisions/branches. Organizational names used here are those that existed at the time of this research.[1]

In many ways, the organizational issues observed in this study remain, because some of the key players who have an important role in developing suppliers lie outside the direct reporting chain of command of AFGLSC or even the Air Force. As we shall see, because the business of key suppliers spans multiple buying units of spares and repairs, including even ALCs, SRM was originally organized outside the normal chains of command and overlaid on existing organizations so that it could take an enterprise-wide perspective. This also complicated its ability to implement strategies in contracts.

The first generation of SRM team leaders, i.e., SSRMs, that we interviewed for this study has turned over and new leaders are now in place. SRM was first established in the Air Force at AFMC in FY 2002 as part of the PSCM initiative, which has played a large role in how SRM was developed and rolled out. The goals of each of the original five SRM teams combine with those of PSCM, which in turn link to eLog21 goals.[2]

AFMC is one of several MAJCOMs within the Air Force. AFMC's subordinate organizations are primarily the product centers where new weapon systems are acquired and the ALCs where fielded systems are supported. AFMC also has technology labs and other specialized units. Secondary items (reparables and their repairs) are managed by the CCs; their contracts and spend data record information at the NSN level. DLA manages consumable spares and,

1 In June 2010, AFMC reorganized to directorates, divisions, and branches that require no minimum number of personnel in each, unlike the standardized, larger-sized wings, groups, and squadrons for other major commands (MAJCOMs) of the Air Force (AFMC Public Affairs, 2010).

2 The original five SRM teams now number three, one at each of the three ALCs. The suppliers managed by the former SRM team led by the Director of Contracting, AFMC/PK, have moved to Ogden ALC. The Deputy Director of Supply, AFMC/A4, continues to serve as the AFMC liaison for DLA and organic maintenance.

since FY 2006, writes contracts for new reparable spares.[3] Other contracts for goods and services that are managed by organizations other than the CCs may include services, PBL, CLS, modifications and upgrade contracts, and so forth. These types of contracts typically do not purchase individual NSNs and are often of larger dollar value. Other Air Force spending outside of AFMC occurs primarily at the other MAJCOMs, generally for base-level requirements, such as base operating support.

Managing Supplier Relationships in the Air Force

The Air Force's goals for SRM mirror those of commercial firms: reduce the cost of goods and services provided and improve quality and performance by carefully managing relations with suppliers to the mutual advantage of each.[4] SRM in the Air Force, which operates under the larger PSCM transformation initiative, also employs techniques similar to those used by commercial firms in its efforts to achieve reductions: take an "enterprise" view of purchasing sustainment items, consolidate contracts to take better advantage of the buying power of large annual expenditures, and work with suppliers on a cooperative basis so that mutual advantages accrue.

But the Air Force is not a commercial firm. The goal of a commercial firm is to make a profit and satisfy its customers, while the purpose of the Air Force is to contribute to the national defense as effectively and efficiently as possible. While the Air Force shares some characteristics with commercial firms in how it operates, e.g., buying parts in the marketplace, it differs from commercial firms in important ways. First, it buys finished products rather than assembling them itself.[5] Further, its purchasing practices are guided by an extensive set of federal regulations, which affects its ability to take full advantage of SRM best practices.[6]

How SRM Works in the Air Force

SRM in the Air Force involves a number of actors—including both individuals and organizations. Table 3.1 lists the major ones and their roles.

Though the AFMC organizational structure referred to here was the one that existed at the time of the research and up to June 2010, when it was replaced, the issues described here are still relevant. Figure 3.1 graphically depicts how these actors interface within the SRM framework. SRM teams were originally set up outside of any organization, because the busi-

---

3 Consumable spares are not economical to repair and are removed and replaced. Reparables are parts that are less expensive to repair than to buy new. *Spares* also refers to new reparables purchased to replace those that have been condemned as not economical to repair or to increase numbers of these items in inventory to meet demand surges.

4 Some companies view SRM as reducing the total cost of business with a given supplier, because SRM encompasses all contracts and dollars.

5 Airlines are similar to the Air Force in that they also buy finished products and do not assemble them.

6 Commercial aerospace original equipment manufacturers (OEMs) must follow Federal Aviation Administration regulations and guidelines for manufacturing and repairing parts, but Federal Acquisition Regulations (FARs) and socioeconomic goals do not apply to commercial contracts.

Table 3.1
AFMC SRM Participants and the Roles They Fill

Participant	Role
SSRM	Individual responsible for leading the SRM team and implementing, developing, and managing supplier relationships[a]
SRM team	Group of individuals, typically working full-time, who work with the supplier(s) to improve quality, cost, and performance
SRM suppliers	Companies that supply greatest amount of materiel or services to Air Force by dollars
OEM	Organization that initially produces a piece of equipment, typically an aerospace manufacturer
CC	Cross-functional group that develops and carries out AFMC-wide sustainment commodity sourcing strategies
Contracting officer	Individual with warrant authority to obligate government funds and write, award, modify, manage, and terminate contracts
Combat sustainment wing (CSW)[b]	Sustainment organization that focuses on supporting parts common to multiple aircraft
Supply chain manager (SCM)	Individual who manages items required for weapon systems; SCMs of common items are in CSWs, and Weapon System SCMs of weapon system–unique items are in ASWs
Aircraft sustainment wing (ASW)	Sustainment organization that focuses on supporting entire weapon systems, including weapon system–unique parts
System program manager (SPM)	Individual responsible for the sustainment of fielded weapon systems that are no longer in production
System support manager (SSM)	Individual responsible for the sustainment of fielded systems that are still in production
Weapon system supply chain manager (WS SCM)	See SCM. Individual who manages items required for weapon systems; WS SCMs of weapon system–unique items are in ASWs
System program office(r) (SPO)	Individual or organization responsible for overseeing the acquisition of new system or major upgrade

[a] At the time the report was written, the SSRM function was an additional duty assigned to Directors of Contracting (office symbol: PK).

[b] CSWs have largely been absorbed into the supply chain management groups (SCMGs) under AFGLSC. Whereas CSWs were more decentralized to the ALCs, the SCMG chain of command reports directly to AFGLSC.

ness of key suppliers can span multiple CCs and ALCs.[7] What follows is a discussion of what key elements of the diagram do within the SRM framework.

Supplier/OEM. The supplier (who could also be an OEM) provides a repair service or spare part to a buyer of one of three organizations: a CC, which purchases like kinds of parts; a CSW, which is a sustainment organization that focuses on supporting parts common to multiple aircraft; or an ASW, which is a sustainment organization that focuses on supporting entire weapon systems, including parts unique to specific weapons systems.[8]

---

[7] With the creation of AFGLSC, SRM teams were reorganized under the new center. The SRM liaison relationship between AFMC and DLA and organic maintenance has also been reorganized under AFGLSC.

[8] During the period of this study, contracting officers were matrixed to various wings. That is, their contracting chain of authority came from the contracting directorates, but they also reported to the organizations where they were assigned, e.g., CCs, CSWs, and ASWs. Under AFGLSC, contracting officers reside in three contracting divisions, one at each ALC, and report to the 448th SCMG and supply chain management wing at Oklahoma City ALC. Contracting officers for buying spares have migrated in place and are now part of DLA.

Figure 3.1
Major Participants in Initial Implementation of Air Force SRM

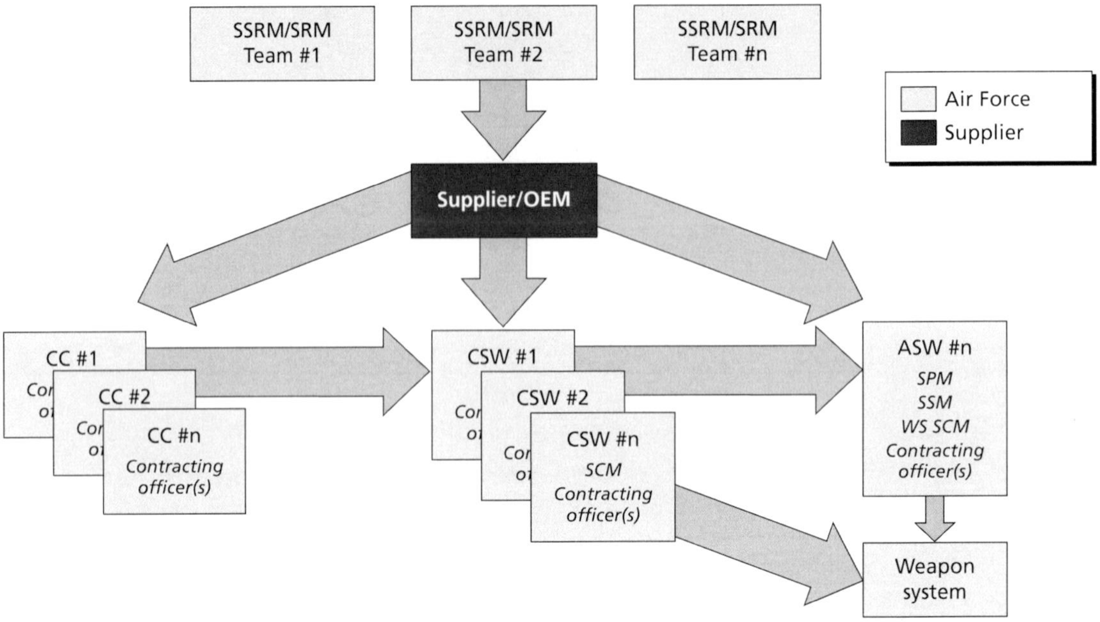

NOTES: Italics indicates individuals, while roman font indicates groups or organizations. Relationships depicted in the diagram are organizational, not necessarily geographical; i.e., entities depicted in the same box may not be in the same physical location.
RAND *TR904-3.1*

The OEMs retain a supplier role once the aircraft has been delivered to the Air Force. Since there are few systems integrator OEMs, the ones that exist today have manufactured more than one aircraft or weapon system. Thus, since they produce and repair both common and weapon system–unique items, they have a relationship with virtually all the stakeholders of the logistics enterprise: those that have a weapon system orientation as well as those with a commodity perspective.

Commodity Council. CCs, as their name implies, have a commodity focus; that is, they focus on similar kinds of items across multiple systems rather than the entire system or even items specific to a particular system. Spend and technical analyses were originally conducted to determine which commodities to group together and where to locate the councils. To maximize their leverage with suppliers, CCs develop sourcing strategies and centrally buy the vast bulk of weapon system secondary items, measured by dollars, that are managed by the CSWs. The CC (of which there were eight in AFMC at the time of this study; we show only three in the figure for illustrative purposes) focuses on a specific group of items, such as propulsion. The focus of the CC is to develop strategic, long-term contracts for most items purchased for the CSWs and, to a lesser extent, the ASWs.

CCs purchase as many of the items needed by the CSWs as possible. Their approach to purchases differs from that of the CSW in that it is strategic, looking across requirements over a long time horizon, rather than tactical, looking at immediate requirements. The CCs make strategic purchases of most of the NSN spares and repairs by leveraging the entire business of selected types of parts. CC supply strategies must be approved by a group of senior leaders made up of the chiefs of the CSWs and the directors of contracting at the ALCs—as the connecting arrows in the diagram indicate. The director of contracting at Headquarters, Air Force Materiel Command (HQ AFMC/PK), chairs this group, referred to later as Tier II of the Logistics Business Board.

Contracting Officer. A contracting officer is an individual who has the authority to obligate government funds and write, award, modify, manage, and terminate contracts. Those located at the CCs, CSWs, and ASWs can contract with suppliers or OEMs for the parts or classes of parts deemed as needed by that organization.

If we look at only those contracts for NSN-level spares and repairs, three organizations write contracts: the CCs, CSWs, and ASWs. As the figure shows, all three organizations have contracting officers, and two, the CSWs and ASWs, have SCMs. As noted above, CCs write contracts primarily to supply parts and repairs for items managed by SCMs in the CSWs and by the WS SCMs in the ASWs. Thus, SCMs in the CSWs and ASWs are stakeholders of the contracts written by the CCs for the items that they manage and that their own contracting officers write.

Combat Sustainment Wing. The CSWs focus on managing groups of parts that are common to multiple systems. They make tactical or short-term purchases of items, either because the items fall outside the purview of the CCs, fail infrequently, or are needed in a hurry. The CSWs also develop purchasing strategies.

Supply Chain Manager. Weapon systems have many unique parts, but they also have many that are common to more than one weapon system. The SCMs in the CSWs manage the common parts used by the organic maintenance shops for subsystem repair and by depots for aircraft in programmed maintenance. Their goal is to achieve the highest possible availability of a given weapon system, subject to resource constraints, by ensuring that the parts necessary to keep their associated weapon systems operating are available when needed. Under SRM, it is important for the SCM to work closely with the CSW (and suppliers) because he or she is an internal customer of the CCs and can influence their supply strategies and contracts as well as the tactical purchases of parts within the CSW.

Aircraft Sustainment Wing. ASWs take a weapon system perspective; i.e., they focus on a specific weapon system, such as F-16 aircraft. Their objective parallels that of the SCMs: optimum availability of a weapon system. ASWs have WS SCMs who are responsible for ensuring that parts and services are available for depot repair operations to meet weapon system availability objectives. They recommend resource allocation and prioritize constraints. They rely on the CSWs for common parts and purchase weapon system–unique items and services.

Within an ASW are Aircraft Sustainment Groups (ASGs) that are responsible for single large weapon systems or similar smaller systems. They write large contracts for programmed depot maintenance or weapon system PBL. SCMs in the ASWs manage the weapon system–unique parts used also by the organic maintenance shops for subsystem repair and by aircraft in programmed depot maintenance. The chief of the ASG is the SPM of a weapon system that is out of production; the SPM is responsible for the weapon system's overall sustainment. SPMs work closely with the SCMs and the suppliers. While a system is still in production, the person responsible for the aircraft's overall sustainment of fielded aircraft is the SSM.

AFMC SRM Practices

As part of PSCM, the same data that were used to define the eight CCs were also used to identify AFMC's top spares and repair suppliers or those suppliers with which the ALCs were conducting the most business.[9] In the early days of SRM, total component spend for spares and repairs with all suppliers was rank-ordered for the period 2002 to 2004, and the top 21 suppliers were assigned to SRM. (In the analyses that appear later in this chapter, these 21 private-sector suppliers are called "SRM suppliers.") Shortly thereafter, at the urging of SRM suppliers and because of their importance as key suppliers of parts and reparables, DLA and ALC organic maintenance were also added as suppliers to be managed and developed under the SRM initiative.[10] A separate liaison relationship was established with senior leadership at HQ AFMC/A4, the directorate of logistics that oversees supply and maintenance, to manage these. Five SRM teams were originally created, one at each ALC, located at Hill, Tinker, and Robins Air Force Bases, and two at HQ AFMC.[11] The locations, number of personnel assigned to each team and their career backgrounds, and the private-sector suppliers assigned to teams appear in Table 3.2.

Senior Executive Service (SES) managers were assigned as managers (SSRMs) of the SRM teams. We were told that this was meant to telegraph the importance of SRM to the suppliers and to internal stakeholders. Various Directors of Contracting throughout AFMC took on the role as SSRM as a logical function, having already established close contacts and direct interactions with contractors. They also participate in the board that approves CC supply strategies described below. At the time of this research, SRM was only a small part of each of the SES SSRM's responsibilities.

The SRM teams are relatively small, but they generally utilize personnel who are dedicated full-time to the SRM effort. The teams are composed primarily of mid-level General Schedule (GS) civilian employees, usually GS-12 or GS-13, who have the backgrounds sought by the SSRMs. These backgrounds vary across teams. Each team managing private-sector suppliers also has at least one hired consultant who brings technical skills and private industry expertise needed by the team, including knowledge of best practices in SRM.[12]

At AFMC, at the time of this study, SRM was the responsibility of two teams. The AFMC/A4 SRM team consisted of an SES and a GS-13 employee and was responsible for DLA

---

9 SRM suppliers were identified by their Contractor and Government Entity (CAGE) codes and scrubbed iteratively with the contractors themselves. CAGE codes are assigned to each operating location of the enterprise by the government and are not as carefully maintained as those of the Data Universal Numbering System (DUNS), which uses nine-digit numbers that uniquely identify individual businesses and can be used to access a database that details the name of the company assigned. Parent companies and their associated "children" businesses were determined from their associated DUNS numbers.

10 DLA manages the bulk of consumable parts (parts that are replaced and not repaired) used as replacements of other parts or by the repair process itself. BRAC 2005 law also assigns DLA the responsibility of contracting for all new reparables that had previously been purchased by the respective military services. These contracting personnel "migrated in place" to DLA and its chain of command, still located at the ALCs.

BRAC law does not apply to reparables purchased on a performance-based contract. The services, and not DLA, will continue to write contracts for these items. In this case, the contractor is held accountable to meet performance goals; supply functions of inventory management, maintenance, and logistics are included as functions to trade off.

11 As noted earlier, the HQ AFMC team of private-sector suppliers has since been combined with the SRM team at Ogden ALC, and responsibility for creating a liaison relationship that communicates AFMC requirements to and works with DLA and the organic maintenance depots has moved from HQ AFMC to HQ AFGLSC.

12 Some consultants were retired Air Force personnel given additional training by their new employer.

Table 3.2
Original AFMC SRM Teams

Teams	HQ AFMC[a]	HQ AFMC[a]	Oklahoma City ALC	Ogden ALC	Warner Robins ALC
No. persons – office symbol	1 – A4 (SSRM) 1 – staff	1 – PK (SSRM) 2 – staff	1 – PK (SSRM) 3 – staff	1 – PK (SSRM) 6 – staff	1– PK (SSRM) 2 – staff
Career field		Contracting; logistic mgt specialist	Logistics; engineering; maint scheduler	Contracting; logistics; stud/contracting	Contracting; acquisitions
Consultants		2	1	1	1
Total (including SSRM)	2	5	5	8	4
Suppliers	DLA[b] Organic maintenance[c]	Raytheon Boeing Northrop Grumman	Pratt & Whitney General Electric Smiths Aerospace Kelly Aviation Parker Hannifin Hamilton Sundstrand Rolls-Royce Chromalloy	Honeywell Goodrich	Lockheed Martin BAE Systems Rockwell Collins EDO Corp. DRS Technologies L-3 Communications Teledyne Crane Aerospace (Signal Technologies)
Total suppliers	2	3	8	2	8
Total SRM spend ($M), FY2006–2008	7,552	950[d]	2,094	369[d]	1,110

[a] The SRM teams at HQ AFMC no longer exist. DLA and organic maintenance are managed by an SRM team located at HQ AFGLSC, and Raytheon, Boeing, and Northrop Grumman are now assigned to OO-ALC.

[b] DLA spend = $2,731 million. SOURCES: FY 2006–2008 DD350 and FPDS-NG data; all DoD spend for SRM CAGE codes.

[c] Organic maintenance = $4,820 million. SOURCE (excluding DLA): FY 2006–2008 NSN-Level Spend for SRM CAGE Codes, Global Combat Support System–Air Force (GCSS-AF) Strategic Sourcing Analysis Tool.

[d] Raytheon, Boeing, and Northrop Grumman are now managed by the OO-ALC SRM team, making the total spend it manages about $1,319 million.

and organic maintenance. The AFMC/PK team was responsible for three private suppliers, in part to give headquarters firsthand experience with the same SRM work that the ALCs conduct (as opposed to taking charge of DLA and organic exclusively). The AFMC/PK team was led by an SES, with two GS-13 employees and two consultants.

The SRM teams located at three ALCs were responsible for 18 of the 21 private suppliers.[13] At Tinker Air Force Base (OC-ALC), the SRM team consisted of the SES SSRM, three civilians (GS-13), and a consultant and was responsible for eight private suppliers. At Hill Air Force Base (OO-ALC), the SRM team had an SES SSRM, six civilian employees (GS-12 to GS-13 and a GS-5 intern), and a consultant and was responsible for two private suppliers.[14] At Robins Air Force Base (WR-ALC), the team consisted of the SES SSRM, two civilians, and a consultant and was responsible for eight private suppliers.

---

[13] Today, the three ALCs manage all 21 private-sector suppliers.

[14] Today, that number has increased to five private suppliers, after the AFMC/PK suppliers were assigned to Ogden ALC.

Also shown in Table 3.2, the number and spend of suppliers relative to the size of the team varied greatly across the sites. The ratio of dedicated staff (excluding SSRMs) to suppliers ranged from approximately one staff member for three suppliers at the lowest to four staff members for one supplier at the highest. The ratio of dedicated staff (excluding SSRMs and consultants) to spend was even wider, ranging from a low of one staff member with DLA and depot maintenance per $61.5 million to a high of one such staff member per $7,552 million; among the private-sector suppliers, the highest ratio was one dedicated staff member per $698 million.

In conducting its supplier development activities, AFMC's SRM organization leveraged the organizational structure set up by the CCs. PSCM had established three tiers of management, referred to as Tier I, II, or III, of the Logistics Business Board that would also serve as the governing structure for developing and approving SRM improvement initiatives and strategies. Tier I, led by the AFMC commander, AFMC/CC, with the headquarters' director of contracting, AFMC/PK, as a sitting member, sets the overall policy direction of PSCM to fit with the long-term vision of the Air Force. Tier II, led by AFMC/A4 and AFMC/PK with participation of the ALC directors of contracting and chiefs of the CSWs, approves commodity acquisition management proposals (CAMPs) for various commodity groups. Tier III, led by CC leaders, develops supply strategies for their respective CAMPs. These proposals are two-year plans for how to purchase groups of goods and services. Included are the acquisition strategies to be used, such as contract type and contract duration. As we shall see, not all suppliers can be managed through CCs, because, as we indicated at the outset of this chapter, the business of key suppliers can span multiple CCs and ALCs.

One of the first accomplishments of AFMC's SRM initiative was to establish an objective baseline of contractor past performance to identify targets of opportunities for improvements. Recall from Chapter Two that a key tenet of SRM is to identify those areas that promise greatest benefit to the enterprise if dealt with and improved. Establishing a baseline requires a way to assess how the supplier is performing across the portfolio of business conducted for the enterprise. This measure of past performance, called a supplier scorecard, helps identify where improvements are needed.

The earliest supplier scorecard metrics included mission-capable (MICAP) hours,[15] customer wait time,[16] procurement lead time, on-time delivery, quality, and deficiency reports (Pfeiffer, 2006). The Air Force later added a cost metric, the Cost Lead Time Demand Index.[17] At initial supplier summits, football team names were substituted for actual supplier names to obscure their identities, but eventually this practice was dropped in favor of reporting scores

---

15 MICAP hours are the hours that equipment is grounded or cannot be operated because of the lack of a serviceable part to replace the one removed.

16 Customer wait time is the elapsed time between an order for a part and when that part is issued from base supply to the customer.

17 Early in the initiative, the cost metric derived by the Air Force could not be populated by data from some of the standard data systems. This issue has since been resolved, according to discussions with AFMC personnel in June 2007. This metric shows the tradeoff among price, cycle times, and inventory levels. It can indicate when reduced cycle times—and higher prices—can lead to lower inventory levels and lower total costs.

directly by supplier. Much of the initial efforts of the SRM teams were spent on developing a consensus on the scorecard metrics and methods for populating them.[18]

SRM suppliers were presented their scores and allowed to dispute the numbers. According to one SSRM, the value of analyzing government databases is to derive an objective value that, even if it is only 80 percent accurate, can still indicate where problems exist that need to be addressed. Moreover, these supplier scorecards are not used to determine contract awards or penalties.[19]

Concurrently, DLA was holding its strategic supplier alliance summits for its largest sole-source suppliers. Supplier representatives and even some Air Force personnel found themselves invited to both summits discussing similar issues. To streamline these efforts and save time and money, AFMC accepted DLA's invitation to cosponsor supplier summits and hold them jointly. AFMC held three supplier summits on its own before joining with DLA for joint supplier summits.[20]

The supplier scorecards and other SRM team analyses helped identify those areas needing overall improvements. AFMC's mechanism for working on specific improvements that can apply to multiple contracts is the Joint Improvement Initiative (JII). These supplier-specific initiatives are intended to be short in duration—one year to 18 months—and have good prospects of benefiting both parties. These JIIs are decided upon through a process of negotiation. They can differ in the scope of what they strive to achieve. Some have bottom-line objectives. Others are more process-oriented and qualitative in nature. Some focus more on improvements that the supplier can make, and others focus on what AFMC can do to improve its processes to lower the cost of competition for third suppliers, such as simplifying the language used in announcing new business opportunities. Some JIIs are suggested by AFMC and others by suppliers. One SRM team relied on analysis to prioritize its JIIs; others, while also conducting analysis, seemed to rely on negotiation to rank their JIIs.

AFMC SRM Outcomes

The wide variation in SRM teams captured in Table 3.2 suggests that we might observe variation in the development of SRM by each team and associated outcomes. Sites with more staff relative to the numbers of suppliers or to supplier spend appeared to have more complex JIIs,

---

18 Many Air Force data systems are legacy systems designed years before relational databases were available. The same kinds of data for a single event, such as product delivery, that are entered into different systems can have different dates and quantities and ultimately lead to different statistics. This can be true when comparing events across Air Force systems and even between Air Force and contractor systems. In some cases, secondary data, such as summaries, derived from original data are retained longer than the original data, making disputes over separate transactions and secondary data difficult to reconcile. For example, DCMA's Mechanization of Contract Administration Services system does not retain history data on actual performance past a certain period, though it does retain performance summaries. Discrepancies over actual performance history are not easily resolved.

19 Resistance to supplier scorecards populated by Air Force data, which can be inaccurate, was explained as a desire to protect the company's reputation.

20 Those we interviewed voiced concerns that joint supplier summits sent the wrong message. A common complaint was the lack of "actionable" items to emerge from these meetings. We were told that, over time, as the Air Force was not in complete control of the agenda, suppliers sent lower-level personnel to these summits. It was also complex to have DLA as a cosponsor of these summits while also representing one of the of the Air Force's key SRM suppliers. We have no evidence of any inhibitions to discussing problems candidly about DLA as a supplier, though the concern is plausible.

engaged in more data analysis, and reported more measurable effects, at least in terms of lead time reductions. The suppliers with the largest spend were assigned to SSRMs with few resources to devote to them. Not surprisingly, less SRM headway appeared to have been made among them, at least in terms of measurable performance improvements or cost reductions. The team that managed government suppliers spent most of its time communicating AFMC's objectives and goals and developing pilot programs for metrics and performance improvements. These relationships are more complex than those with private-sector suppliers. One SSRM of a small team noted that it was hard to justify additional resources without having measurable results to show for the resources already spent, but it is possible that this team simply had too few resources to have effected any measurable changes in the time allotted. Finally, we note that these results likely reflect other variations across sites as well (e.g., different approaches to SRM); however, staff ratios seem to be a major factor driving the number and complexity of JIIs.

CHAPTER FOUR

Challenges and Hindrances to Implementing Supplier Relationship Management in Air Force Materiel Command

We now turn to issues that hinder AFMC's ability to make greater progress in managing and helping its key sustainment suppliers to improve their costs, quality, and performance. We begin with an analysis of the scope of spend relevant to SRM, which affects the incentives commercial firms have to participate actively in the SRM process. Next, we identify specific aspects of the Air Force SRM initiative and its implementation that impose barriers to a broader or more efficient implementation of SRM in the Air Force.

SRM teams were given general guidance but wide discretion to develop practices and processes used to manage their key suppliers. SRM grew naturally from the PSCM initiative that transformed the purchasing and supply chain management processes for spares and repair. But it is now challenging for SRM teams to make greater progress, especially with their larger customers, such as Lockheed Martin, Boeing, Northrop Grumman, and Raytheon. The willingness of these corporations to invest resources in improvements may be tied to the business base included in SRM, which as we will see, is relatively small. Other challenges include an absence of clear Air Force and DoD policies on SRM, the resources devoted to the program, the capabilities of the personnel assigned to the SRM initiative, and the process of getting SRM concepts implemented in contracts.

The Portion of Spend Considered in SRM Limits Potential Benefits

Our analyses of Air Force spend covered in AFMC's SRM efforts indicate that the narrow scope of contracts considered limits the benefits that the Air Force can hope to achieve. There are two aspects to this, which we address in turn in this section:

- SRM covers only a small portion of the Air Force's total spend.
- For most SRM suppliers, only a portion of their business with the Air Force is covered by SRM.

In addition, our analyses of Air Force spend data reinforce the need for SRM, due to the complexity of the Air Force's business with the SRM suppliers.

SRM Covers Only a Small Portion of the Air Force's Total Spend

The spend that SRM teams manage constitutes only a small part of the Air Force's total contract spend. This limited scope restricts the magnitude of potential savings from SRM.

Considering only one year, FY 2008, Figure 4.1 dissects the Air Force's total budget in a number of ways, placing SRM spend in the broader context of Air Force contracts. Air Force SRM includes only $6.5 billion of a total budget of $160 billion.[1] About 29 percent of the Air Force's budget for FY 2008 is associated with military and civilian personnel costs; nonpersonnel costs make up the remaining 71 percent, or about $114 billion. Over 55 percent ($64 billion) of the Air Force's nonpersonnel budget is attributable to direct spend, i.e., contracts with private-sector companies.[2]

AFMC spending measured by contract spending is almost 71 percent ($45 billion) of the Air Force's direct spend portion of the budget in FY 2008; this includes new weapon systems, upgrades and modernization programs, and depot-level spending. The ALCs constitute 20 percent of the Air Force's direct spend portion of the budget ($13 billion). Finally, SRM private-sector supplier spending—both NSN-level and non-NSN-level—makes up 10 percent of the Air Force's direct spend.[3] The potential effect of Air Force SRM efforts relates directly to the portion of spending to which it is applied and is low.

Figure 4.1
SRM Share of FY 2008 Air Force Budget

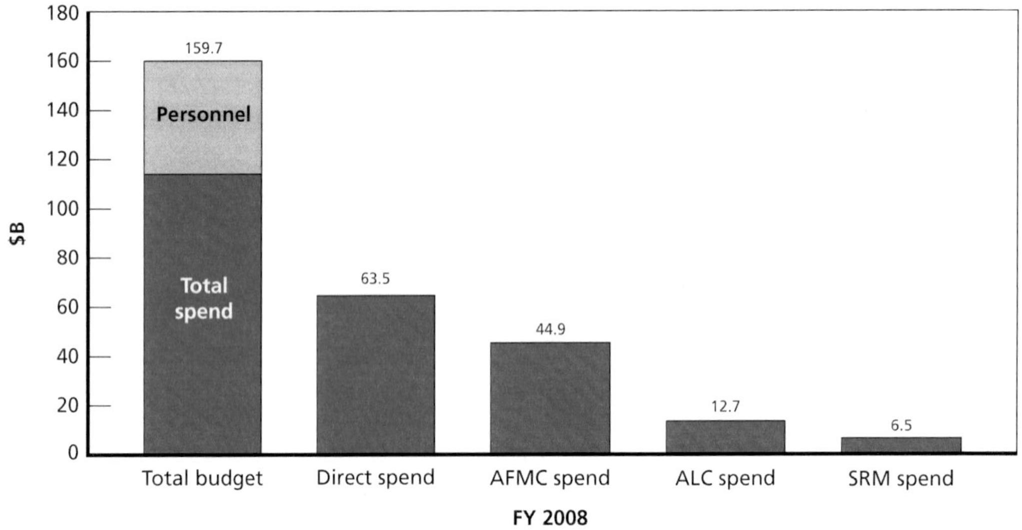

SOURCES: For total budget: Office of the Under Secretary of Defense (Comptroller), 2009, Tables 6–18. For all others: U.S. Department of Defense, Defense Manpower Data Center, Statistical Information Analysis Division, no date; AFMC-provided SRM CAGE codes.
RAND *TR904-4.1*

---

[1] Note that this $6.5 billion, which includes all ALC spending for the 21 SRM suppliers, should not be confused with the almost $6.0 billion in Table B.1, which was a three-year total for 21 suppliers for only NSN-level spending. The $6.5 billion estimates the ceiling of what could be included in SRM in FY 2008 if all ALC spend for these 21 suppliers were considered.

[2] "Nondirect" spending is anything that is not personnel and not found in the DD350 or FPDS-NG data that is reported as an Air Force purchase. Examples include purchases made on behalf of the Air Force by DLA, USTRANSCOM, General Services Administration, Army, Navy, Missile Defense Agency, etc., as well as certain classified contracts.

[3] This estimate of 10 percent of the Air Force's total spend is an upper bound for all business of these SRM suppliers at the ALCs, more than the NSN-level spend that AFMC considers for SRM.

For Most SRM Suppliers, Only a Portion of Their Business with the Air Force Is Covered by SRM

Even within the small portion of Air Force direct spend covered by SRM, benefits are limited by the fact that not all of the SRM suppliers' spend is covered in the program, reducing incentives for suppliers to participate fully in initiatives to reduce costs and increase performance.

If most of a company's business were included in the SRM initiative, the incentives for responding to Air Force initiatives would be high.[4] The converse should also hold true—that is, if very little of the company's business is included in SRM, the company may make a perfunctory effort to respond to Air Force initiatives for improvement but be less likely to invest time or financial resources. The individual program manager at each company may be very motivated, but he or she must compete with other corporate activities to devote personnel and financial resources to improvement exercises.

Figure 4.2 shows the total Air Force spend with the 21 SRM suppliers during fiscal years 2006–2008, broken down among categories.[5] The dark blue sections illustrate the percentage of spend with each SRM supplier that is included in the SRM program. Firms are ordered from top to bottom by decreasing percentage of Air Force spend covered by SRM.

Most of the contractors that have the majority of their direct Air Force business included in the SRM initiative are Tier 1 contractors. These firms often subcontract with prime contractors for production of new weapon systems. As examples, virtually all of Parker Hannifan Corporation's and Crane Aerospace and Electronics' (Signal Technologies) direct Air Force business is included in the SRM initiative.

For over half of the SRM suppliers, most of their Air Force business lies outside of AFMC's SRM initiative. For some, such as Lockheed Martin, Boeing, L-3 Communications, and Northrop Grumman, very little of their Air Force business—less than 5 percent—is covered by the SRM initiative.

To put spending among the SRM suppliers in perspective, the right-hand side of Figure 4.2 shows the percentage of Air Force spend among this group of 21 suppliers that each contractor constitutes. For example, in FY 2006–2008, the Air Force spent 40 percent of its contract dollars associated with SRM suppliers with Lockheed Martin, 23 percent of with Boeing, and 13 percent with Northrop Grumman. Just these three companies made up 76 percent of the Air Force's total direct spend for SRM suppliers. On the other hand, the eight contractors

[4] We believe this would be the case even for sole-source suppliers. Cooperation with SRM could boost customer satisfaction, reduce costs, and increase the profit on fixed-price contracts. Over time, as contractors pass savings through to the Air Force, prices should decrease or, for the same price performance, quality should increase.

[5] The DD350 data include spares and repair contract dollars. To derive total spending with each SRM contractor, we "overlaid" spares and repair contract dollars on ALC spending. What remained became "other ALC" spending. Similar logic was used for the other categories of spending. To derive total Air Force spend, we took the total dollars provided by FY 2006 DD350 and FY 2007–2008 FPDS-NG data and combined them with the SRM NSN-level data to show the portion of the budget that is managed by SRM. DD350 data for FY 2006 are available from the Office of the Secretary of Defense. (See U.S. Department of Defense, Defense Manpower Data Center, Statistical Information Analysis Division, 2008.) Before FY 2005, transactions of at least $25,000 were required to be recorded on DD350 forms. Beginning in FY 2005, FAR 4.602(c)(1) required that transactions of at least $2,500 be recorded on DD350 forms. Contract actions of at least $25,000 were recorded in FY 2004 and actions of at least $2,500 were recorded in FY 2005 and FY 2006. Most Air Force dollars are captured even with the higher threshold. FPDS-NG data for FY 2007 and 2008 are available online from the Federal Procurement Data System website (Federal Procurement Data System, no date).

Figure 4.2
Air Force Spend with SRM Suppliers, FY 2006–2008

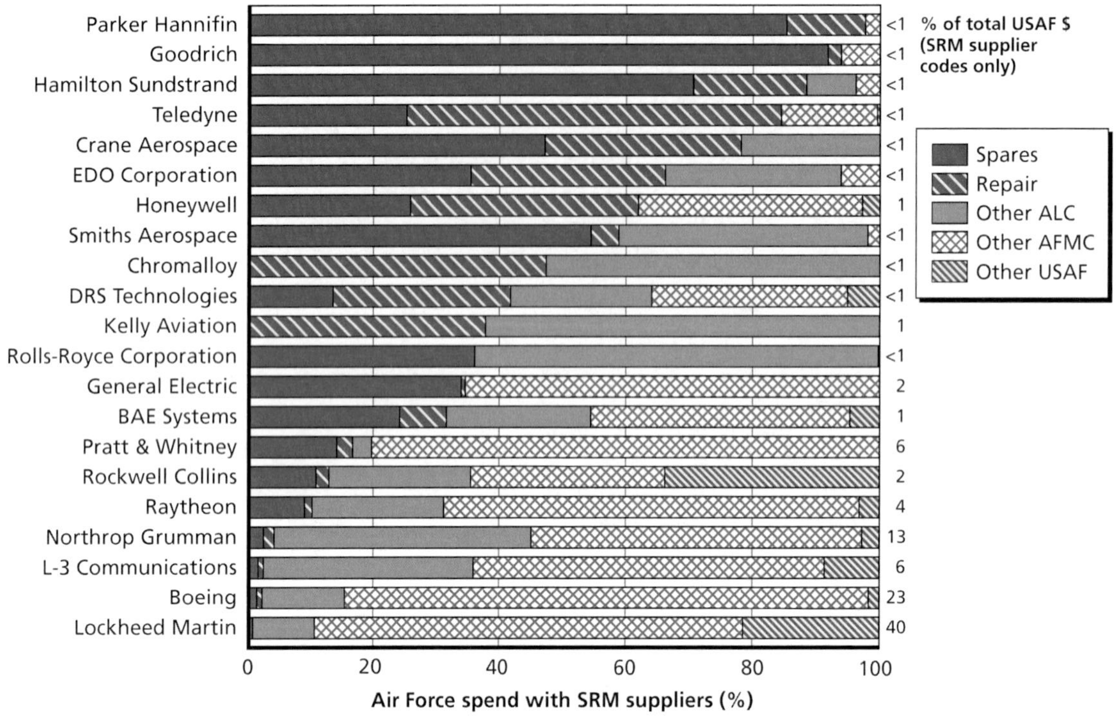

SOURCE: FY 2006-2008 DD350 and FPDS-NG data, NSN-Level Contract Data (Strategic Sourcing Analysis Tool), and SRM supplier codes assigned by DoD, called Commercial and Government Entity codes.
NOTE: Percentages may not sum to 100 due to rounding.
RAND *TR904-4.2*

with the highest percentage of business in the SRM initiative made up less than 2 percent of the Air Force's dollars.

If SRM were to expand to include all of ALC spend, it would include more of the Air Force's direct spend with contractors such as Northrop Grumman, L-3 Communications, Raytheon, and BAE Systems. Only if SRM were to expand to other parts of AFMC, such as the rest of the spend at the ALCs and its weapon system–buying organizations, the product centers, would the initiative encompass most of the business base of all SRM suppliers, including the large OEMs. As noted above, without more comprehensive inclusion of these firms' business with the Air Force, their incentive to fully invest in and implement SRM improvement initiatives is low.

Most SRM Supplier Spend Is Spread Across Commodity Councils, Increasing the Complexity of SRM Implementation

In this section, we assess how SRM supplier spend is distributed across the CCs within AFMC. Figure 4.3 portrays sustainment spending on spares and repair purchases broken out by CC during FY 2006–2008. The horizontal bars indicate the percentage of total spares and repair (or NSN-level) business for each of the 21 SRM suppliers associated with the CCs at the three ALCs. Shown from left to right is spending by the CCs at OC-ALC, OO-ALC, and WR-

Figure 4.3
AFMC Spending by SRM Supplier and Commodity Council, Spares and Repair, FY 2006–2008

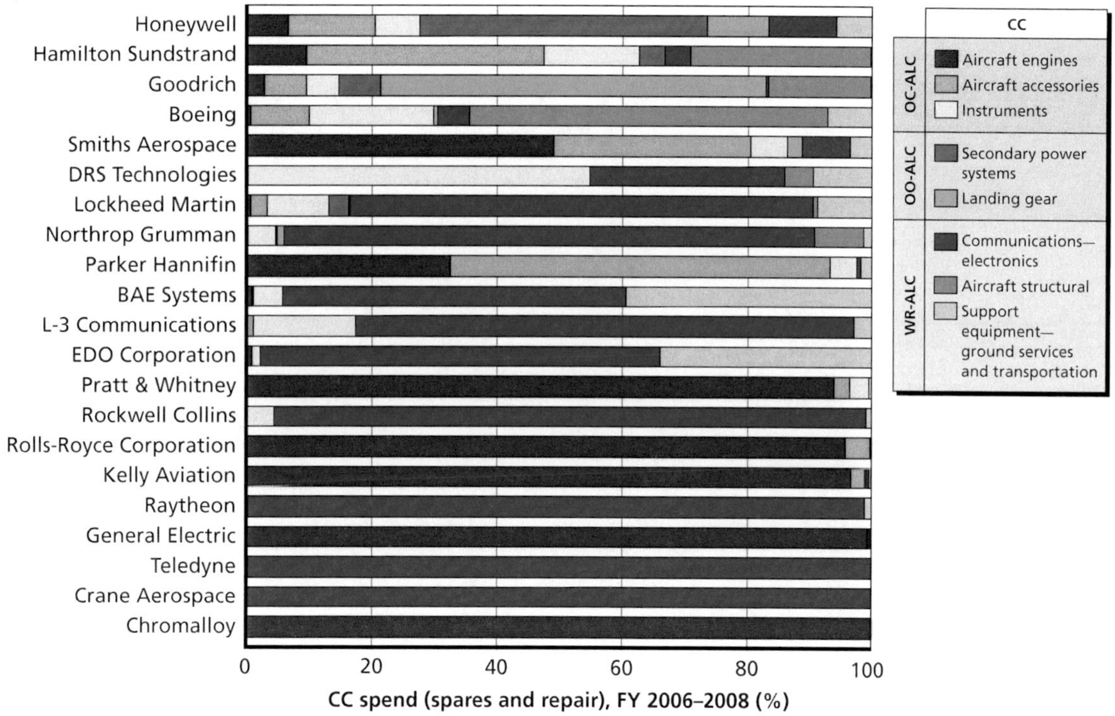

SOURCE: FY 2006–2008 NSN-level spend data for SRM CAGE codes and CC Federal Supply Classes (FSC)s, Strategic Sourcing Analysis Tool.
RAND *TR904-4.3*

ALC.[6] Suppliers are listed in descending order by the extent of spend distributed across the CCs.

The greater the distribution of spending across CCs, the greater the need for an overarching SRM program that can look across the Air Force's business with suppliers to take a comprehensive approach to supplier development. This dispersion increases the complexity of implementing SRM initiatives, since there are more Air Force organizations that must be involved. As examples, the spares and repair business for Honeywell, Hamilton Sundstrand, Goodrich, and Boeing are spread broadly across the ALCs and CCs.

Alternatively, for suppliers with more concentrated spending within a single CC at a single location, SRM teams can work closely with those councils to include development and implementation of JIIs. Considering only spares and repair spending, SRM suppliers Rolls-Royce, Kelly Aviation, Raytheon, General Electric, Teledyne, Crane Aerospace (Signal Technologies), and Chromalloy fall within this category, since their spend is heavily concentrated in a single CC.

This concentration of spend to a single CC as observed here is due in large part to the type of spend managed by SRM. If SRM were expanded to include more than spares and repair spending, supplier management could not be done by a single CC and would, in fact, require

6 Supplier spend was identified by CAGE codes as provided by AFMC.

a separate group or team that could look across all supplier spending and business units when considering initiatives to improve supplier quality, cost, and performance.

Remaining analyses conducted on SRM spend across several different dimensions are contained in Appendix B. For example, Appendix B contains analyses of the total spend of key suppliers managed by SRM in FY 2006–2008, the distribution of SRM spend across the eight CCs for NSN-level repair, only, and the business base included in SRM compared with the total business base of the SRM suppliers for all of DoD.

Overall Observations on Spending and Suppliers

Spend included in SRM constitutes less than 5 percent of the Air Force's budget and about 10 percent of the direct spending. SRM initiatives can only reduce costs and save money within the amount of spending that it influences, so if the Air Force wants to see greater return from SRM, it needs to expand the spend base. Below, we discuss some ways the Air Force might expand the SRM spending base.

The analyses of the supplier NSN-level spending indicate which suppliers the Air Force should cultivate for sustainment improvements. The Air Force has identified its top 21 suppliers, but there are interesting subgroups even within this group. One involves those suppliers that contract across the ALCs and the CCs. These include Honeywell, Goodrich, Boeing, and Hamilton Sundstrand. In theory, it would be difficult to improve the overall performance of these suppliers without an initiative like SRM. At the same time, improving their performance offers the potential for greater savings.

A second group involves those whose spares and repair purchases primarily involve a single CC. Such companies are Rolls-Royce, Kelly Aviation, Pratt & Whitney, Raytheon, General Electric, Crane Aerospace (Signal Technologies), and Chromalloy. It would be inefficient to attempt to manage these suppliers much above the CC level if only NSN-level spend were considered.

A third group of interest contains those suppliers for which Air Force spending constitutes a large fraction of their business base. These companies have an inherent incentive to work with the Air Force, because its spending is important to their bottom lines and because any steps that mutually benefit the supplier and the Air Force will increase their profit margin.

SRM Team Challenges and Hindrances

As noted above, the relatively small spend in and of itself is a barrier to garnering full benefits of an SRM program. However, other aspects also can inhibit more effective and efficient implementation of SRM. Some of these pertain to the SRM teams themselves, and we discuss them here.

Our analysis, based on well-established theory and group and individual interviews with SRM teams and their members, indicates that three general characteristics are necessary for successful SRM implementation.[7] First, a program needs clearly defined goals. These provide

[7] See Appendix C for a discussion of the SRM interviews and Appendix D for the government and private-sector interview protocols.

focus and direction and indicate the tasks and responsibilities needed to accomplish the goals. With goals in hand, a program next needs resources to execute the various responsibilities, usually in the form of physical or human capital. Finally, programs need not just people, but people with the requisite capabilities. Such people ensure that resources are used wisely and efficiently to meet the program's goals; team members know what needs to be done, and they have the means and the skills with which to do it. In addition, sustained top management support is required to ensure that these three elements are in place.

As we see in this section, the Air Force faces challenges in each of these areas that affect its ability to successfully implement SRM.

SRM Implementation in AFMC Lacks Clearly Defined Goals

Our analysis found that Air Force SRM teams lack clearly defined SRM goals and responsibilities. Evidence for this comes from three findings: inability of SRM personnel to articulate clear goals, resistance of other Air Force personnel to implementing initiatives, and the nature of the SRM training.

First, many personnel when asked could not articulate either the goals of SRM or their own responsibilities with any specificity. They articulated broad overarching goals, such as improving purchasing by building relationships. However, few could state the purpose of SRM in terms of definable, measurable goals either for the Air Force in general or for their own jobs in particular. In fact, most SRM personnel did not have a clear or concise definition of their SRM job but rather described it as more self-defined. Some of this is because some tasks vary from supplier to supplier, but, when probed further, personnel's responses appeared to reflect imprecision about the job at hand much more than variability across suppliers. As two staff members put it, "I don't like to flounder; I like to know what my responsibilities are, what's expected of me," and "If you're not a self-starter, [this job] could be really hard."

Compounding the problem of unclear objectives and responsibilities was variation in the frequency and level of SSRM involvement in SRM (e.g., meetings with staff, interactions with suppliers and with internal partners, translating SRM vision into priorities). SRM team members reported that their interactions with their SSRM ranged from every day to every few weeks. Notably, those personnel on SRM teams with reportedly more SSRM involvement expressed a much clearer understanding of their goals and responsibilities.[8] For example, one team member who described regular weekly meetings with the SSRM had a ready goal for SRM—build relationships with the suppliers—and a purpose behind that goal:

> Because that makes things flow better in the tactical world. Making that company relationship better . . . makes the suppliers more responsive to the tactical needs, and then they are more willing to supply [the tactical personnel] with what they need.

Not surprisingly, personnel who reported that their SSRMs were strongly committed to SRM expressed more commitment themselves. Similarly, personnel who reported little interac-

---

[8] Differing degrees of meeting frequencies between SRM teams and their SSRMs and different methods of communication seemed to affect the teams' understanding of goals and what was needed to achieve them. Some SRM teams met with their SSRM weekly and discussed progress on each supplier and JII; another team communicated with its SSRM via regular written reports. This SSRM cited the value of these reports as a means of facilitating needed communication and guidance in the face of real-time constraints. Teams that met with their SSRM irregularly expressed the least commitment to SRM as an important goal for the Air Force.

tion with their SSRM and little involvement of their SSRM in the supplier relationships also showed the least enthusiasm for SRM themselves. In addition, those who could see the results of their work expressed greater belief in the efficacy of SRM (although the causality here could easily run in both directions).

Stakeholder Issues

This section describes stakeholder issues that SRM teams told us complicated their ability to implement JIIs. The process and relationships among stakeholders described here were what existed at the time of this research.

SRM teams require the assistance of multiple internal stakeholders, and executing SRM initiatives requires CCs or SCMs in particular. The second piece of evidence for some lack of clearly defined goals and responsibilities was the obstacles that SRM team members encountered from other Air Force stakeholder personnel, such as SCMs, who did not recognize their role in SRM or the importance of SRM objectives in their jobs. As we described in Chapter Two, CCs have contracting officers who are needed to execute some SRM concepts, and SCMs have contracting officers on staff to purchase items not handled by the CCs. Whereas the CCs focus more on strategic purchasing, the SCMs reportedly focus on tactical buys or buys made in response to actual demands. SCMs are customers of the CCs and their chain of command; the chiefs of the CSWs are part of the Tier II board that approves CC supply strategies. In either case, asking CCs and SCMs to implement JIIs increased the workload for their already fully loaded staff and can require already approved supply strategies to be reopened for further negotiation and administration.

SRM teams described mixed results in working with these two groups. Almost all of the teams characterized relations with CCs as constructive. This is particularly good news given that most JIIs are likely implemented by CCs. Most of the contracting officers are located in the CCs, and for suppliers that span the councils and ALCs, if CCs concur with proposed JIIs, new contracts for these suppliers should incorporate SRM innovations. As long as SRM is focused on NSN-level purchases, getting council agreement on implementing concepts has a positive effect.

One team we interviewed described a sometimes less cooperative atmosphere with their councils, saying "Our commodity councils are kind of territorial," and "there's always the sensitivity of who gets credit for what." However, the following comment was more typical for most teams: "The commodity council is very good. . . . The commodity council is like us—they're transformational, so they're willing because that's what they're tasked to do."

While CCs are charged with transformation, SSRMs also appear to play a direct and important role in supporting positive relationships between the councils and SRM teams. One SRM team member described it this way, "[The SSRM] has walked into [the CCs'] offices and said, 'We have to work together.' [The SSRM] works very hard to try to integrate what I call a transitional work team" to proactively build relationships with internal stakeholders, not just with SRM suppliers. SRM teams also described the contributions that they make to enable their relationships with the councils: "The commodity council has one contracting officer, so it overwhelms them. So it is a team effort—who really sits down and writes the contract."

In addition, on a policy level, the CCs produce commodity acquisition management plans or supply strategies that must be approved by Tier II managers, who include the SSRMs. To implement a JII might mean revisiting these plans, and this would mean backtracking on already completed work for the council and Tier II managers. Since the SSRMs are also Tier

II managers, there is an understanding of why this work may be necessary. Taken together, a spirit of mutual understanding and effort appears to exist between CCs and SRM much of the time. This cooperation may be because the SSRMs fill a dual role across the two groups, because the SRM teams are providing what may be perceived to be free labor, or because the SSRMs have otherwise engaged the CCs to participate.

In contrast to the relatively good relationships with CCs, SRM personnel described trying to get assistance from SCMs as one of their biggest challenges. Repeatedly, we heard that persons within the Air Force, i.e., those who would be colleagues in a supplier negotiation, were far more difficult to work with than were the suppliers. When asked who these internal stakeholders were, we heard that SCMs required more time to bring onboard. One staff member said, "Actually, the suppliers are very willing; they're out in the general world always wanting ways to improve business. The commodity council is very good; it's the supply chain managers who have a little more resistance." SCMs are responsible for tactical purchasing that responds to immediate supply needs. SRM teams' requests for assistance in streamlining these contracts, reassessing plans that SCMs view as ready for execution, or implementing JIIs interrupts this real-time process. The lack of measurable benefits of SRM also provides an opportunity for those who question the benefit of these efforts to resist incorporating SRM concepts into new contracts. In addition, SCMs support SPMs, who may not be knowledgeable about the prospective benefits of SRM. As stakeholders at the weapon system level, SPMs wield considerable influence, since they are accountable for aircraft performance. It would require a far-thinking SPM to be willing to expend scarce resources for long-term gains, which may not be realized until he or she has moved on and which compete with other needs.

In light of resource constraints (i.e., a heavy workload), SCMs understandably might not heed requests for assistance unless they understood the importance of the SRM. Communicating the importance of JIIs and persuading SCMs to assist with their implementation appears to be the responsibility of SRM team members. While this may be a reasonable division of labor, SRM team members appear compelled to spend an inordinate amount of time working with other stakeholders. One staff member put it this way: "The supplier relationship in some ways was much easier, because we're the customer. Most of them really want to work with us; they're an easier sell than the government people." Some of this resistance no doubt derives from the individual's particular role and perspective within the organization. For SRM to work, some of these inherent challenges will need to be better understood and resolved.

The third observation that indicates a lack of clarity about SRM goals and responsibilities comes from what SRM personnel reported about SRM training. Standardized SRM training appears to be minimal, consisting of a module on SRM within the two-week PSCM immersion training package that is usually available only to the SSRM. Most personnel reported that they had not received this training, and those who did report receiving it said that on-the-job training, which is open-ended, was more useful than one short (i.e., a few hours) module. However, on-the-job training also depends on the support of the SRM team providing that training. For example, one staff member reported, "All of my training has been from the team and on the job, doing it. . . . I've been tickled working with these guys, it's a good group," while another stated, "This is not the center focus of our business; it's a side focus, so I don't see a big training effort." As a result, the definition of SRM and the approach used vary across the sites. This variation might not be a problem if the teams' efforts could be measured and if their effectiveness were consistent. However, responsibilities are often not clearly enough defined to measure. Moreover, training in SRM should be available to the CSW and ASW SCMs.

SRM Resources Might Be Inadequate

As with any wide-reaching new program, identifying and implementing ways of managing and developing suppliers requires substantial staff time. First, personnel are ideally dedicated solely to SRM duties, i.e., not splitting time with other responsibilities, so that they may focus on SRM and be most efficient. Second, there need to be enough personnel assigned to complete the tasks at hand. For new programs in particular, the number of personnel also needs to be assessed regularly to make sure the number remains sufficient. Third, the number and spend of suppliers relative to the size of the SRM team affects the speed of putting SRM into practice as well as its effect. This is likely not a linear relationship; rather, a minimum number of SRM team members are needed for a particular number of suppliers or a particular amount of spend before any progress can be expected at all, after which additional numbers of personnel should enable further progress.

SRM staffing results have been mixed. First, we observed that almost all of the teams (not including SSRMs) were filled with personnel whose sole responsibility was SRM. One exception was the headquarters teams, whose members also had other responsibilities. On other teams, the only diversion from SRM activities was time that personnel had to devote to short-term tasks, such as symposium planning. This is a notable accomplishment of the Air Force, because it reflects an understanding of the needs of SRM and sends the message that SRM cannot be administered effectively on the side.

Second, in terms of numbers of staff, we observed wide variation across the sites, with the size of SRM teams (including the SSRM) ranging from two to eight people, with most teams having five or fewer SRM staff. Thus, the amount of work that the SRM teams can accomplish with suppliers is necessarily constrained. The larger teams appeared to have more complex JIIs focused on measurable cost reduction and involving extensive data analysis. Their team members also reported more interactions with suppliers and internal partners and covered staff departures more effectively by distributing the departing person's work across more team members. When asked how they had been able to accomplish a particular goal, one team member summarized the situation this way: "We make it happen through research and relationship." In contrast, teams with fewer staff appeared far more limited in their efforts. One team member expressed the difference between what needed to be done and what there were resources to do this way:

> I think the missed opportunity is the data mining on the scorecard. I think there was too much focus on the grade. So if they're reporting red on something, you really need to sit down with that contractor and say, "Hey, what's going on here?" We really need to do the drill-down to see if there's a problem.

Most SSRMs and their team members did note the potential usefulness of having more personnel than they currently had. However, most did not see increasing their staff size as possible—some vacant SRM slots had remained unfilled for some time (i.e., after a team member was promoted), effectively shrinking the size of the team. Most SSRMs and SRM staff reported that higher priorities prevented them from reasonably arguing for more staff. The way one person put it was as follows:

> I think that's the only logical place [for funding], because it is a PSCM, CSW initiative. It's a logical thing that they would be the source for resources, funding, and slots. Unless you

wanted to get the commander involved, which you wouldn't for something like this. . . . They were not convinced that it was worth it to provide more effort to these contractors. . . . We reasoned we just had to make the best with the resources we had.

We were told that OO-ALC had a large SRM staff because the SSRM, who had earlier been a leader in the PSCM initiative, went to the ALC commander and HQ AFMC/PK and negotiated more staff. In general, however, the limited number of personnel slots devoted to SRM as well as the image at most sites of SRM as having a lower priority indicate that the resource requirements of SRM—and the benefits of SRM—either are not fully understood at AFMC or fall victim to higher priorities. Fully instituted, SRM catalyzes a sea change in the way the Air Force purchases sustainment goods and services, and it needs corresponding resources to accomplish this, at least in the program's start-up years.

Our above observations on staffing adequacies were based on interviews with SRM teams and the organizational structure that existed at the time of the research. Analyses that measure workloads, required skills, and personnel-to-workload ratios are recommended to estimate staffing adequacy.

SRM Needs Certain Personnel Qualities and Skills

In addition to clearly defined goals and sufficient resources, transformational initiatives such as SRM need capable people with the right mix of skills to execute them effectively. The right mix of skills can boost an organization's ability to manage change successfully, which can affect the speed of transformation. A successful program requires goals and responsibilities, resources, and particular skills. These skills may be individual aptitudes or they may be gained through experience or training. According to our interviewees, abilities relevant to SRM fall into the areas of interpersonal skills, data analysis, and innovative thinking, as shown in Table 4.1. Some of these are used more in the early stages of building and developing the program (innovative thinking and certain interpersonal skills, such as marketing), while others are needed both then and after SRM is fully instituted (interpersonal skills and data analysis).

Virtually all SRM staff reported that interpersonal skills were essential for their job. Indeed, they are fundamental to relationship building. SRM personnel relied on their ability

Table 4.1
Individual Skills Useful for Supplier Relationship Management

Skill	Characteristics
Interpersonal	Politeness and respect
	Empathy
	Persistence
	Marketing, persuasion
Analytic	Data procurement
	Data preparation
	Quantitative data analysis
Innovative thinking	Understanding new approaches
	Developing new methods for new approaches
	Being a self-starter

to interact with people and to build relationships in all aspects of their jobs. They obtained supplier involvement for developing and implementing initiatives, engaged internal stakeholders as needed for the initiatives, and worked together as a team, drawing on members' experience in multiple career fields. SRM personnel also identified several particular interpersonal skills that facilitated their work. Persistence was often needed to gain internal stakeholders' participation, because SRM was not necessarily among their highest priorities and because they sometimes resisted the changes that SRM brought. Similarly, many team members reported that they wished they had training in marketing because they used it so often to explain SRM to internal stakeholders as well as suppliers and to "sell" them on it.

Data analysis skills are necessary for analyzing sustainment and supplier data to develop JIIs. Analyses of data can reveal suppliers' strongest and weakest areas, thereby pointing out where to focus SRM efforts. They can also suggest areas that suppliers' SRM teams should focus on more heavily for particular measures, especially relative to their spend. This process is necessary to forming JIIs that show measurable effects. SRM teams need both supplier data and DCMA data from the CCs, and some SRM personnel reported having a hard time getting portions of these data. In addition, many personnel reported that cleaning the data, i.e., making them reliable for analysis, took concerted effort. Thus, the analytic skills needed include data procurement, data preparation, and data analysis. While all of the teams recognized the importance of data analysis, more measurable JIIs that utilize those skills are needed for SRM to be instituted fully.

Innovative thinking is another skill important for SRM, especially its development. Because putting SRM into practice involves nothing less than a sea change for managing and developing Air Force sustainment suppliers, SRM personnel themselves need to be able to think in new ways. Innovative thinking enables the team members to grasp SRM and identify which practices can be most effective at addressing priority needs, and it enables them to come up with ways to get others to adopt new practices. This is important for their relationships with both suppliers and internal stakeholders.

Related to this, many SRM personnel (including SSRMs) also identified being a self-starter or independent worker as an important skill. This skill may be at least partly an adaptation to not having the goals and responsibilities required by SRM clearly defined. However, the strategic nature of the work also suggests that the ability to plan and implement one's responsibilities independently is a definitive part of SRM. In addition, the strategic nature of SRM certainly differs from the mostly tactical experience of many of the personnel.

Overall, the Air Force appears to have done an impressive job identifying and hiring persons for SRM teams who possess most or all of these skills. With few exceptions, the teams appear to be staffed largely by people with strong interpersonal, analytic, and innovative skills. We note, however, that the small number of personnel on these teams, ranging from two to eight including the SSRMs, make it imperative for members to draw expertise from outside their immediate teams to bring a cross-functional perspective to supplier development efforts.

AFMC SRM Implementation Needs Sustained Top Management Support

Top management support—through both resources and management attention—is a critical success factor for successful implementation of SRM. While SRM enjoyed early support from top management at the highest levels, with the commander of AFMC attending early supplier summits, that support needs to continue to visibly communicate to Air Force and supplier organizations the importance of these efforts. Such sustained, highly visible support can be conveyed through leadership setting the agenda of supplier summits, periodic meetings between top Air Force leadership and top supplier management, sharing SRM-related metrics broadly with both Air Force and supplier personnel, and creating a sense of importance that SRM succeed. Without this kind of support, SRM faces organizational resistance and has less of a chance to succeed.

Other Factors Inhibiting SRM Implementation

In addition to the relatively small spend and some inherent aspects of the SRM teams, other factors work to limit the effectiveness of SRM in the Air Force, and we describe these next.

SRM Has Been Overlaid on an Existing Organizational Structure

SRM is not as tightly woven into the organizational fabric as is the case with commercial SRM organizations. Rather than being designed into the organization, it has been overlaid on the existing Air Force logistical structure. This means that the SSRM must insert his or her organization into processes that were not necessarily designed to accommodate SRM. In practical terms, this means that he or she must use a business case put together by the SRM team to persuade other organizations of the advantages of integrating SRM practices into ongoing operations and contracts. An example of such a business case might be the anticipated improvements of on-time deliveries as a result of sharing forecast data with suppliers as soon as such data become available. These other organizations ultimately must agree with the SRM teams before supplier improvement ideas developed by these teams are implemented into new contracts, so getting their "buy-in" is critical. Also, we note that the duty of the SSRM is an additional one assigned to directors of contracting. This means that, at a minimum, the SSRM must allocate some of his or her attention to many other competing responsibilities. One result, discussed below, is that SRM implementation and the role that the SSRM plays varies across the Air Force.

Multiple Players with Overlapping Roles Create Coordination Challenges

For SRM to achieve the goal of working with suppliers to reduce costs and improve quality and performance, the initiatives developed by the SRM team and each supplier must be implemented across all units doing business with a supplier. SRM needs to be integrated into all contracting and requirements organizations if it is to succeed in furthering higher-level, enterprise-wide goals, such as eLog21. Initiatives that pertain only to process changes typically can occur without a contract. However, most initiatives pertaining to how the contractor performs the work and the kind of product or service delivered require implementation within contracts. Thus, for SRM to succeed, supplier-specific initiatives will need to be implemented in most, or even all, of the sustainment contracts for that supplier. This becomes challenging for suppliers working across multiple organizations, as seen in Figure 4.3, particularly for suppliers also working with non–Air Force organizations, such as DLA.

More broadly, as the discussion of Figure 3.1 in Chapter Three suggested, the logistics system has many actors, and several of them have overlapping roles. The many players make coordination of supplier initiatives complex, even for relatively straightforward tasks. As noted above, SRM team members report that much of their time is spent marketing initiatives to the various stakeholders of the contracts associated with the supplier they are managing. Having so many stakeholders means that any one of them can prevent SRM initiatives from being implemented in that part of the business. Commercial companies have addressed this issue by adding new business that a supplier wins to existing contracts and managing the entire business of a supplier on few contracts or even a single contract. This option is not readily available to the government.

A Career Development Plan Is Lacking for SRM Personnel

As noted in Chapter Two, firms that have successfully implemented SRM select personnel who have the right aptitude and "soft" skills and train them on any needed technical expertise and skills. As a symptom of the organizational challenges discussed above, the Air Force lacks a plan to identify and develop personnel who have the capability to undertake supplier development activities. In addition, within the current personnel system, it is unclear whether SRM experience helps or hurts personnel's chances for future advancement. In our interviews, personnel who perceived SRM as not advantageous for their future career advancement also expressed less commitment to the program.

Current Metrics Do Not Support Aggressive Implementation of SRM

Leading commercial enterprises use scorecards both as literal measures of how suppliers are performing and as indicators where and how much they need to improve. Their suppliers, often fiercely protective of brands and reputations, will work hard to raise low scores or maintain good ones. Supplier scorecards are also a succinct means of communicating to the buying enterprise how its supply base is performing, and scorecards are sometimes a key factor when awarding new or renewed business. The government has limitations on how these measures might be used for determining new contract awards, but supplier concerns over scorecards using these final numbers suggest that suppliers, including DLA and organic maintenance, care deeply about them and could be motivated to improve them. In fact, at the start of AFMC's SRM effort, considerable time was initially spent working with SRM suppliers to review and validate inputs used to derive metrics.

Similar metrics can be used to motivate SRM team performance. Members of SRM teams reported that metrics for SRM personnel were considered early on to motivate team members to succeed, but were largely rejected because even though personnel might have been given the responsibility for helping suppliers improve, they had limited authority over most of the processes necessary for bringing those improvements to fruition. One team considered using a quantifiable goal but set a modest goal instead, for the reasons stated above.

Similarly, some teams set quantifiable goals for suppliers to achieve, but they were generally modest. Though the idea was to provide a real target of improvement, teams were concerned that a metric set too high would discourage suppliers from agreeing to improve their performance. The most common metric and improvement initiatives were on-time deliveries.

Summary

To summarize, personnel need a clear understanding of the importance of SRM to the enterprise, what it is, and how those outside of the SRM teams are expected to support SRM and JIIs. SRM team members reported that initiatives sometimes received lukewarm support from other units concerned that SRM initiatives would actually harm existing support, even where supplier performance problems existed, and units were reluctant to get behind ideas if a convincing business case could not be made. Personnel performance metrics for players outside of the SRM teams also tend not to be tied to supplier performance, which means that incentives for personnel who might be important to supplier improvement plans are weakly linked to their success or failure. As with many large organizations, new initiatives can face strong resistance unless their goals and objectives are articulated and readily understood, the initiatives are perceived to be backed by top management, and all players share incentives to support the initiative.

CHAPTER FIVE

Conclusions and Recommendations

We now turn to our final conclusions about how the Air Force has implemented SRM best practices and about what hinders even more thorough implementation. We also offer recommendations to increase the effectiveness of the SRM initiative at AFMC.

Conclusions

Our research found that AFMC has incorporated some SRM best practices in its program. Focusing primarily on NSN-level spend of the eight CCs, AFMC identified top suppliers by dollars and assigned them to SRM teams at the ALC where most SRM business took place. Three large SRM suppliers—Northrop Grumman, Boeing, and Raytheon—were the exception; however, they are now managed by teams located at an ALC. Two government organizations supplying the largest supply of parts were included as well in SRM—the organic depots and DLA. Supplier scorecards were developed to show how suppliers were performing overall and in detail. Supplier summits are used to bring together both Air Force and supplier senior leadership to talk about goals and objectives and act as a clearinghouse for solutions to various problems. JIIs that are mutually agreed on by the SRM teams and suppliers form the basis of developing improvements in processes and outcomes. SRM team members are usually full-time employees who have been selected on the basis of their skills for self-initiation, powers of persuasion, and hard work. All of these practices mirror what best-in-class companies have developed for their own SRM programs.

That noted, our research also identified several major factors that seem to be keeping the Air Force's SRM program from accomplishing more. First and most importantly, the SRM program's current scope includes too narrow a slice of Air Force and supplier spend. As a result, it limits supplier incentives to participate and the potential overall impact for the Air Force. Second, well-defined goals do not appear to be communicated clearly and consistently among SRM teams or within AFMC in general. As a result, for the study period, many SRM staff do not appear to understand their responsibilities completely, and other internal stakeholders do not appear to understand the purpose of SRM. Third, there do not appear to be enough personnel resources, given the number of suppliers and level of spend at most sites. Fourth, the program lacks the sustained senior leader support needed to institutionalize SRM. Finally, the way in which SRM integrates into other AFMC initiatives creates organizational complexity that hinders effective and efficient implementation. The commitment of SRM personnel to their job may erode without strong SSRM leadership, evidence of the benefit of SRM, and support to maintain their functional career field development.

Recommendations

Our analysis of SRM best practices and Air Force implementation leads us to several recommendations. We group these recommendations into three categories: (1) expanding the business base of SRM, (2) institutionalizing SRM, and (3) developing needed analytic capabilities to support supplier development.

Expand the Business Base

A serious weakness of SRM is that its current business base is restricted to spares and repair purchases, which is a small portion of the business base that most of these suppliers have with AFMC and the Air Force as a whole. It might be tempting to locate SRM with CCs where, in some cases, much of the SRM spend is associated with a single CC; however, as our analyses have shown, this base represents only a small part of the business base that some of the largest suppliers have with the Air Force. The paradox is that to meet PSCM and eLog21 goals, AFMC needs to see improvements from its top sustainment suppliers who are also OEMs, yet much of AFMC's expenditures with them is not yet included in SRM.

To enhance supplier incentives to improve, we recommend that SRM be expanded to cover more of the total business that these suppliers have with the Air Force. In particular, SRM should expand to cover all business with the ALCs as well as the product centers. One means of linking the sustainment business base (the little "a") to the acquisition business base (the big "A") would be to develop an integrated supplier scorecard for the whole of AFMC. Another means might be for the product centers to develop their own SRM initiative to operate above the program level—according to whatever the laws permit—and act as a clearinghouse for identifying common issues that, if addressed by the supplier in one program, could benefit those in others. We are aware of several suppliers that have this type of SRM structure within their organizations—production and sustainment SRM programs that share information and use an integrated supplier scorecard. Still another means might be to establish a joint product center–ALC SRM program in which the total business base of key suppliers is included with both types of businesses (big "A" and little "a") represented.

Institutionalize SRM

Institutionalizing SRM to make it an integral set of practices used for managing as much of the business of all key suppliers as possible would provide it more influence and direction. Steps toward institutionalizing SRM include the following:

- sustained support from top management, which would communicate that top management values SRM and intends to cascade those values throughout the enterprise and supply chain
- formal SRM policies and guidance that clarify the objectives, roles, and responsibilities of SRM teams and key stakeholders
- official recognition of the duties of SRM for SSRMs
- formal training programs
- cross-functional support available to SRM teams
- process changes that facilitate implementation of SRM initiatives across organizations and locations

- methods of measuring and sharing metrics of supplier performance improvements across the enterprise.

Unless such steps are taken, SRM risks becoming marginalized and losing resources, importance, and, inevitably, influence. This would then adversely impact other related initiatives, such as eLog21 and supply chain risk management.

One hallmark of institutionalizing SRM would be to raise the importance of SRM within the organization. Companies that have seriously implemented SRM consider it a core capability for how they work with their key suppliers. Their SRM teams tend to be small (except for production operations), but they have the strong backing of senior management. The SRM role is well defined, and the rules for working with others in the organization are well understood. Less time is spent on "marketing" and more on working directly with suppliers to improve. Team members tend to be analytical and well trained, with clear career incentives and development paths. Concepts are quickly implemented, often by groups outside of SRM.

To ensure that SRM becomes an organic part of the organization, AFMC should develop further policy and guidance that describes the roles and responsibilities of SRM teams, stakeholders, and key suppliers, as well as metrics and processes for developing JIIs to assist SRM teams in implementing best practices in a systematic way. Developing policy and guidance and making SRM permanent would help to secure resources and staffing for these activities in the long- and short-term budgeting cycles. AFMC should also make SRM an official responsibility of the SSRMs within their portfolio of responsibilities. This would help the SSRMs rank SRM as one of their priorities and provide them incentives to devote more time to it.

AFMC should also consider SRM in its career development of those types of individuals who could be recruited onto SRM teams. To maintain momentum and the direction of SRM and assist with career development, AFMC should institute formal SRM training, such as a stand-alone training course specific to SRM, and require all stakeholders to be trained on what SRM is, why the Air Force needs it, and what constitutes best practices. This would lift some of the burden from SRM teams for educating stakeholders about SRM. AFMC will need to work with DLA on training its buyers on AFMC's SRM initiative as part of their career development. These buyers will play an integral part of implementing concepts developed by SRM teams for reparable spares. Adequacy of staffing should also be assessed. SRM teams should, optimally, operate as separate teams and be adequately staffed and able to look across the AFMC enterprise and total supplier business base.

To institutionalize SRM, AFMC must also consider how to facilitate the implementation of SRM concepts in contracts and revised processes. Rather than working purely by consensus, where just one stakeholder can prevent the implementation of good but new ideas, AFMC should implement a standardized approval process for improvement initiatives similar to that of supply strategies for the CCs. To bring improvement initiatives into the CC supply strategy formation process would require thinking about the interface between SRM teams and the councils and how it could be improved. Since the CCs are the buyers of goods and services, incentives should exist to consider SRM concepts and, if they are promising, to support them. After all, these councils will directly benefit from supplier improvements. According to our interviews, shoehorning SRM concepts into already-approved supply strategies has hindered JII implementation. In addition, if supplier spend concentrates heavily in one CC, it might make sense for the SRM team to work predominantly just with that council. An enterprise-wide spend analysis would indicate which suppliers fall into this category. Finally, an approval process is also needed to implement SRM concepts within contracts that fall outside the purview of the CCs. Contracts for weapon system unique items and non-NSN level spending fall into this category.

Increasing SRM influence and institutionalizing it also opens the door to becoming more serious about expectations of supplier improvements. AFMC should move toward developing clearly defined, aggressive targets that are shared and supported by top management so that the goals are easily and well understood throughout all levels of AFMC and key supplier enterprises. These targets also provide useful feedback on whether supplier improvements are moving in the intended direction. This means coming up with quantifiable targets for what needs to be improved and how much. Metrics that measure the performance of SRM teams and all relevant stakeholders and link to supplier improvements would also send a message that the success of SRM is important to the enterprise and requires the assistance of personnel other than just the SRM team.

Develop Analytic Capabilities

Finally, AFMC should continue to develop an analytical capability that can help populate supplier scorecards and identify and justify targets of opportunity for improvements, including development of business cases needed to secure support from top Air Force management, various stakeholders, suppliers, and DLA and organic maintenance. SRM should be able to benefit directly from the Air Force's new enterprise resource planning system, the Expeditionary Combat Support System (ECSS), because it should facilitate the collection of data needed for SRM-related analyses. SRM teams will also need to keep pace with best practices. The companies we interviewed have made this capability organic to the SRM team.

SRM has grown in importance in the private sector over recent years. AFMC's experience with SRM suggests that while it has tried to implement many of the best practices, some key hindrances prevent it from reaping real benefits from the initiative. If addressed, SRM should become a significant tool for AFMC in seeking quality, cost, and performance improvements with its key suppliers.

APPENDIX A

Four Stages of Krause and Handfield Methodology

In this appendix, we describe in detail the steps within the four stages of the Krause and Handfield (1999) supplier development methodology, along with the associated critical success factors.

Stage 1: Identify, Assess, and Rationalize the Supply Base

The goal of Stage 1 is to develop a pool of suppliers that are potentially capable of meeting the buying enterprise's supply requirements. Stage 1 has four steps:

1. **Identify strategic supply chain needs,** which are driven by customer demands for dramatic performance improvements in quality, cost, delivery/cycle time, and/or technology and financial health, management, and capacity. The top ten supplier improvement goals identified by their survey in order of their ranking were (1) improved quality of purchased item or service, (2) reduced *total* price of the purchased item or service, (3) improved delivery responsiveness, (4) reduced *unit* price of the purchased item or service, (5) increased supplier service/responsiveness, (6) improved supplier technical capability, (7) increased product development capability of the supplier, (8) increased managerial capability of the supplier, (9) reduction of firm's supply base, and (10) increased financial strength of the supplier.
2. **Search for competitive suppliers worldwide.** Global economics (e.g., risk from fluctuating exchange rates), intensified global competition, and domestic content laws and other protectionist actions of some countries are driving enterprises to seek global suppliers for their operations.
   - 2a. *Focused search in target country or region.* Once a decision has been made to seek suppliers in a specific area, buying enterprises use various strategies to identify prospective suppliers there, such as hosting local conferences and meetings, establishing an International Procurement Office (IPO) in the area, and working with local government development agencies, which may also provide local training.
   - 2a. *Narrow supply options.* Once contact has been made with local suppliers, those that clearly do not meet buyer requirements as outlined in the buying enter-

prises supplier policy[1] and commodity/category strategy need to be eliminated from consideration, while those with the potential to meet buyer requirements receive further review.

3. **Establish performance metrics and assess suppliers.** Many leading enterprises develop a performance measurement system to assess and track their suppliers' performance over time to identify quickly any supplier problems that need immediate attention; target areas and suppliers for supplier development efforts; encourage supplier performance improvement; and identify poorly performing suppliers that will not or cannot improve to phase out, as well as high-performing suppliers for more business. For example, Altel issues scorecards to its 35 strategic suppliers,[2] which represent 65 percent of its spending. Ideally, these performance measurement systems should be real-time and provide suppliers with immediate feedback, but many are less automated and performed quarterly. The specific metrics enterprises use to measure supplier performance vary and often depend on their customers' and their strategic goals, but most include some measure of quality, price/cost/total cost, delivery, and cycle time, as well as other important metrics to the enterprise, such as product and process technology, engineering capabilities, and management skills.[3] Leading companies also develop a formal process for communicating with suppliers and providing feedback on their performance.[4]
4. **Rationalize the supply base.** Because suppliers' development efforts can require significant financial and personnel resources, eliminating suppliers that are clearly not capable of meeting current and future needs and concentrating businesses with fewer, better-performing, sometimes global suppliers is an important prerequisite for supplier development. A number of leading enterprises that have already significantly reduced their supply base report that their rationalization efforts are ongoing as they continually seek to replace poorer performing and/or less capable suppliers with better performing and/or more capable suppliers.

Stage 2: Problem-Solving Development

In Stage 2, buyers identify supplier capabilities, strengths, and weaknesses/deficiencies; establish a baseline of suppliers' core processes and outputs; and, where appropriate, work with suppliers to correct any deficiencies in current performance. The goal of Stage 2 is a supply base

---

1 See, for example, Honda of America Manufacturing Inc.'s web page on "How to Become a Supplier," which lays out Honda's expectations for suppliers (Honda, no date).

2 Strategic suppliers are ones that are embedded in Altel's operations, and switching business to another supplier would involve very high costs. They are also companies that Altel believes will, if it spent time to invest in the relationship with them, have positive results because they bring a differential or competitive advantage. Thus, networking equipment billing systems or anything that generates revenue is strategic to Altel. See Carbone, 2007.

3 For example, the purpose of Altel's supplier scorecard is to improve communications with its strategic suppliers as well as suppliers' performance. Its scorecard metrics can be different depending on the supplier because its goals and objectives are not always quality, delivery, or cost. Altel believes it is a mistake to try to apply the same scorecard to every commodity and every supplier relationship. See Carbone, 2007.

4 Altel looks for continuous improvement on scorecards. It wants to see quarter-over-quarter improvement with suppliers (Carbone, 2007).

that meets the buying enterprise's current production requirements. This stage contains the following two steps:

5. **On-site risk assessment by cross-functional teams.** The purpose of this step is to obtain a thorough understanding of a supplier's capabilities, strengths, and weaknesses/ deficiencies, as well as its ability to meet future requirements. Buying enterprises often develop a risk assessment measurement protocol that teams use during on-site supplier visits.[5] Typical assessment areas include quality; cycle time; changeover, performance measurement, inventory management, cost-reduction, technical, ramp-up, and project management capabilities; purchasing and sourcing skills; sophistication of information systems; financial health; management vision; and labor and environmental performance. These assessments help suppliers understand the buying enterprise's requirements and are sometimes used to qualify a supplier's processes. One outcome of the assessment may be a continuous improvement plan for the supplier.
6. **Problem-solving to eliminate suppliers' deficiencies ("rcactive").**[6] The purpose of this step, i.e., reactive supplier development, is to fix current supplier problems that affect the buying enterprise's production and require immediate attention. These problems are often identified by the buying enterprise's supplier performance measurement system or its on-site risk assessment team.[7] Supplier problems or deficiencies not related to short-term production needs, such as efficiency and capabilities, are deferred for future improvement efforts.

Stage 3: Proactive Development

The goal of Stage 3 is a self-reliant supply base that continually improves to meet evolving buyer needs. Only about one-third of the companies Krause and Handfield (1999) interviewed had developed their supply bases to this stage, and many had only done so in one particular region. More important, these companies acknowledged that significant further progress could be made by better integrating suppliers through the supply network. Stage 3 consists of the following three steps:

7. **Establish open relationships through feedback and information sharing.** This step sets the stage for proactive supplier development. It begins by targeting suppliers to develop. These are typically suppliers that produce a critical commodity that is high-cost, -value, or -volume, etc.; have a systematic history of problems; and have a management that appears willing to discuss problems. Once suppliers have been selected for proactive development, dialogue is initiated with suppliers' top management to com-

---

[5] Some automotive companies use QS-9000 criteria and some electronics/electrical companies use the Standard Supplier Quality Assessment criteria developed by SEMATECH for their assessment protocol (Krause and Handfield, 1999).

[6] For at least one-third of the companies Krause and Handfield (1999) interviewed, reactive problem-solving was their most advanced level of supplier development.

[7] Krause and Handfield's survey of U.S. purchasing managers (1999) found that supplier development teams were typically cross-functional and most often involved personnel from the purchasing, quality assurance, engineering, materials management, and manufacturing departments (in descending order of participation).

municate the buying enterprise's intent, understand any supplier reluctance to open up its facilities to the buyer to improve its performance, and discuss future actions identified in Steps 3 and 5 to improve supplier efficiency, capabilities, or performance. Ideally, a candid discussion of supplier problems takes place, agreement is reached regarding the need for improvement, and both parties express a willingness to cooperate, get to the root causes of problems, and solve them.[8] The goal of these discussions is to identify a specific process for a JII with a high likelihood for quick success.[9] The outcome of the meeting(s) is often a letter of agreement signed by both companies specifying the time and resources each will commit to the JII.

8. **Systematic supplier development.** While enterprises vary in their approaches to proactive supplier development, they typically use one or more of the following three practices to speed supplier improvements:[10]
   8a. *Direct involvement activities.* Some enterprises take a hands-on approach and send experienced personnel to supplier facilities to help them fix problems and develop capabilities. These efforts may include *kaizen* events, process mapping, work-in-process inventory reductions, total preventative maintenance, and other joint projects as well as education and training for supplier personnel in improvement techniques, such as Statistical Process Control and Pareto Analysis, and in personnel, environmental, and supply base management.[11]
   8b. *Incentives and rewards.* Some enterprises use incentives and rewards, such as providing financial assistance,[12] increasing order volumes, steering more future business to suppliers that improve within a certain time frame, and hosting annual award ceremonies where superior supplier performance is recognized, to encourage their supplies to continually improve.[13] These can also complement the buying enterprise's direct supplier development activities.

---

8 Buying enterprises can contribute to 40 to 60 percent of supplier problems by such actions as producing inaccurate forecasts, changing requirements without notifying suppliers, and making design changes at the last minute (Krause and Handfield, 1999).

9 Joint improvement projects, sometimes called *kaizen* (i.e., continuous, incremental improvement) events, breakthroughs, or blitzes, can last anywhere from one to 24 weeks. Key objectives of the projects are to demonstrate to supplier personnel the feasibility of improvement and tangible results that can be obtained from using a basic continuous improvement model, such as "Plan-Do-Check-Act" cycle, which was introduced by W. Edwards Deming and is sometimes referred to as a Shewhart or Deming Cycle (Krause and Handfield, 1999).

10 Respondents to Krause and Handfield's survey (1999) of U.S. purchasing managers reported that visits by personnel (supplier personnel to buyer and buyer personnel to supplier) were the most extensive direct involvement activities, followed by high-level meetings with supplier upper management, joint cost-saving projects, shared cost savings, and encouraging suppliers to improve management of their suppliers. Less utilized practices, in descending order of utilization, were regular visits by buyer engineering personnel to suppliers' facilities, dedicated supplier development teams, allocation of buyer personnel to improve supplier's technical skills base, training/education of supplier personnel, and direct investment in supplier operations.

11 Supplier development efforts can be particularly challenging when supplier personnel turnover is high.

12 Financial incentives, which are most likely to be used in regions experiencing economic duress, include paying supplier invoices early, providing access to loans at lower interest rates, and investing in supplier facilities and equipment (Krause and Handfield, 1999).

13 U.S. purchasing managers reported that the promise of future business if performance improves was the most used supplier incentive, followed by promise of additional current volume if performance improves (unlikely for single-source suppliers), and awards recognizing supplier performance improvements (Krause and Handfield, 1999).

8c. *Warnings and penalties.* Some enterprises use negative incentives to indicate to suppliers the consequences of continuing poor performance and/or failure to improve.[14] These include withholding future business or introducing competition by putting the work out for bid or using multiple suppliers for the same good or service to encourage suppliers to improve their performance.

9. **Maintain momentum.** Without proper reinforcements, there is a danger that lessons learned during initial proactive supplier development efforts will be lost and suppliers will return to their old behavior and practices. Ideally, buying enterprises want suppliers to maintain high performance and initiate improvements projects throughout their operations on their own. Many institutionalize the use of measurement-driven supplier performance metrics, incentives and rewards, and warnings and penalties (particularly the fear of losing business) discussed above in their supplier relationships to reinforce supplier development efforts and motivate suppliers to improve further. Another way some buying enterprises try to prevent supplier from backsliding, maintain the initial momentum of their supplier development efforts, and help suppliers become self-reliant is to follow up with monthly supplier visits.

Stage 4: Integrative Development

Stage 4 is the last stage of Krause and Handfield's supplier development model. The goal of Stage 4 is a globally aligned supplier network; this has the potential to deliver the highest level of benefits. Few buying enterprises have made progress in this area of supplier management. Krause and Handfield (1999) note that even the most advanced automotive and electric/electrical enterprises they interviewed had not yet achieved this final level of integration. Stage 4 consists of the following three steps:

10. **Supplier integration in new product/process development.** The purpose of integrating suppliers early in new product/process development is to speed the new product development and production process, incorporate the latest technology into new products and services, and prevent costs from being designed into products,[15] thus making them more competitive.[16] It enables buyers to influence suppliers' new product and process development efforts and suppliers to influence the design of products to improve manufacturability, serviceability, standardization, and quality. Involving suppliers in new product development also enables them to better plan for facilities, capacity, and new production ramp-up, which is particularly important in products with short production life cycles. Suppliers are integrated into new product design through mecha-

---

[14] Krause and Handfield's survey (1999) of U.S. purchasing managers reported that verbal or written requests that a supplier improve performance were most often used followed by putting the item out for bid, or using two or three suppliers for the same good or service. Least reported was using four or more suppliers for the same good or service.

[15] Approximately 70 percent of a product's cost is determined in the design/development phase. See Teague, 2005.

[16] U.S. purchasing managers reported to Krause and Handfield (1999) that supplier involvement in new product development was most often initiated during the product design phase, the next-most-often supplier involvement occurred during the concept phase, followed slightly less frequently by supplier involvement at the production, prototype, and preproduction phases.

nisms such as collocation of supplier personnel at buyer facilities and vice versa, "guest engineers,"[17] sharing technology road maps,[18] and integrating information systems.

11. **Establish performance improvement in second-tier suppliers.** Poor performance by second-tier or lower suppliers can adversely affect supply chain performance. For example, design changes by a second-tier or lower supplier can affect product or process performance, and a lower-tier supplier's poor products, late deliveries, or capacity problems can adversely affect a first-tier supplier's ability to meet the buying enterprise's requirements. Consequently, some buying enterprises are beginning to develop supply chain maps[19] of their "extended enterprise"[20] and performing risk and capability assessments of key lower-tier suppliers so that they can understand their impacts on key product supply chain processes and production capacities. In some cases, buying enterprises will leverage their total buying power with common lower-tier commodity suppliers, such as steel, for their smaller first- and lower-tier suppliers. A few buying enterprises are encouraging and/or helping their first-tier suppliers to better manage their lower-tier suppliers,[21] and some are reaching out and forming relationships with key lower-tier suppliers themselves (Amaral, Billington, and Tsay, 2004; Carbone, 2004b; Goodman, 2006; Carbone, 2001; and Carbone, 2004a). Lastly, some buying enterprises, such as Cisco Systems, are trying to use the Internet to provide visibility, monitor performance, and coordinate plans over more than three tiers of its supply base (Lee, 2002).
12. **Establish an integrated supplier network.** This was the most advanced supplier development strategy being deployed by a few of Krause and Handfield's (1999) case companies. Ideally, these buying enterprises want their suppliers to be able to supply any location worldwide, at a competitive price, with comparable quality, delivery, and technology performance and to participate with them in global growth opportunities. To achieve this goal, buying enterprises are encouraging their best suppliers to grow and develop capabilities in regions where they are developing new production facilities. They are also encouraging their suppliers to set up supplier associations where they share information and educate each other on the latest technologies and best business practices.

---

[17] Guest engineers are supplier engineers involved in product development working alongside the buying enterprise's engineers at R&D facilities. See Krause and Handfield, 1999.

[18] A technology road map is a description over time of how an enterprise plans to incorporate new and emerging technologies into its goods and services and/or operations. Buying enterprises share their product technology road maps with key suppliers to help ensure that suppliers are capable of meeting future technology requirements and that they have the capability before being involved in new product development. See Krause and Handfield, 1999.

[19] For a workbook outlining how to map the supply chain, see Jones and Womack, 2002.

[20] In 1999, Chrysler Corporation (now DaimlerChrysler) was granted a trademark for the term *extended enterprise*. Chrysler's (abbreviated) definition of *extended enterprise* is:

> Extending business relationships by providing process management consultation and workshops to . . . suppliers and supplier tiers in order to reduce cycle time, to minimize system cost, and to improve the quality of the goods or services provided by the suppliers. (Ericksen and Suri, 2001)

[21] Few respondents to Krause and Handfield's survey (1999) of U.S. purchasing managers indicated that they are encouraging their first-tier suppliers to improve the management of their suppliers.

Critical Success Factors

The following five critical success factors are required for successful implementation of Krause and Handfield's supplier development model:

- **Top management support.** As mentioned in Step 7, the support of the buying enterprise's and supplier's top management is important to gain resources, schedules, and formal agreements for unilateral and joint supplier development efforts.
- **Cross-functional support.** As mentioned in Step 5, involvement by a number of different functional personnel from the buying enterprise, including purchasing, quality assurance, engineering, materials management, manufacturing, and other functions, are required to perform supplier risk and other assessments—a necessary step prior to proactive supplier development.
- **Continuous improvement process focus.** A culture of continuous improvement, initially at the buying enterprise, and eventually at supplier enterprises, is critical to the success of the supplier development model and SRM in general. A number of techniques (e.g., Statistical Process Control [SPC];[22] lean, *kaizen* events; Six Sigma;[23] TQM for continuous improvement; and *kaikaku*,[24] process reengineering for discontinuous improvement) have emerged from the business literature to help enterprises design quality into their products and processes, control their processes, and improve them. These techniques help enterprises establish rigorous processes for successfully analyzing problems, identifying root causes, developing preventative actions, and monitoring process outcomes. They also provide a common language for communicating within enterprises across locations, functions, and management levels, as well as among supply chain partners on the need for change and ways to effectively obtain improvements.
- **Global perspective.** A global perspective is especially important for Stage 4, Integrative Development. Competition and the global, interconnected economy are driving enterprises to think and act more globally. This includes growing organically or through mergers and acquisitions to meet global demand; sourcing globally to deliver the best value to customers as well as meet local content laws and aid local product acceptance; and establishing, where appropriate, local production facilities. Some enterprises are requiring suppliers to supply the enterprise's global facilities equally well and measuring suppliers' performance globally.
- **Global information systems.** As enterprises and their supply chains become more globally dispersed, global information systems are required to effectively communicate within and among them. Such systems can communicate forecasts, production requirements, design changes, and forthcoming solicitations. They can also facilitate new product development activities and supplier performance measurement and feedback. Lastly, they can serve as a repository of enterprise knowledge on supply markets (e.g., capacities, prices,

---

[22] SPC is a method of monitoring, controlling, and, ideally, improving a process through statistical analysis. Its four basic steps include measuring the process, eliminating variances in the process to make it consistent, monitoring the process, and improving the process to its best target value.

[23] Six Sigma seeks to identify and remove the causes of defects and errors in manufacturing and business processes. It uses a set of quality management methods, including statistical methods, and creates a special infrastructure of people within the organization (e.g., "Black Belts") who are experts in these methods. See Wikipedia, no date.

[24] *Kaikaku* is the radical improvement of an activity to eliminate waste.

and trends); suppliers (e.g., availability, quality, design capabilities); local laws, regulations, and customs; and competitors.

Many suppliers provide goods or services within a specific industry, which can simplify SRM. For example, Altel, the wireless service provider, notes that in its industry most suppliers are reasonably specific to a commodity (Carbone, 2007). When suppliers produce within an industry or commodity group and buying enterprises have created commodity teams to buy goods and services centrally, these teams can often also manage and develop suppliers for their commodity group, because they have little or no overlap of business with other major commodity groups. For example, Nortel's supplier portfolio teams manage certain product commodities and are responsible for developing strategies and evaluating the supplier's performance (Carbone, 2005b). Sun Microsystem's ten commodity teams are responsible for supplier evaluations (Carbone, 2005a). And Hewlett-Packard's (HP's) commodity teams manage supplier relationships and set supplier strategies (Carbone, 2004b).

However, not all suppliers provide goods within a specific industry or commodity group. Some provide multiple commodities to an enterprise. When that is the case, some buying enterprises create centralized supplier managers to manage these suppliers from an enterprise-wide perspective. For example, IBM production purchasing is developing "enterprise relationships" with suppliers to manage the overall relationship that exists between IBM and a supplier that is providing multiple commodities (Carbone, 1999). At HP, suppliers are often providing multiple commodities to multiple HP business groups. HP has about 45 executive sponsors (vice presidents or senior vice presidents at HP) who meet with suppliers at least twice a year to review business issues. These sponsors are part of HP's strategy to make its relationships with suppliers more synergistic and less transactional. Before the sponsor meets with a supplier, HP purchasers provide an extensive summary of its business with the supplier and categorize spending and opportunities in every commodity (Carvone, 2004b).

APPENDIX B

Supplier Relationship Management Spend Analyses: Methodology and Additional Results

This appendix includes additional spend analyses that we conducted that augment those described in Chapter Four. It includes detail on our spend methodology and additional results along additional dimensions as described in the text.

To analyze total spend, we overlaid the SRM spend with AFMC spending for all spares and repair contract transactions primarily managed by the CCs and the Air Force's total spend, as measured by DD350 and FPDS-NG dollars, i.e., dollars spent on all contract transactions. Figure B.1 shows all of the ALC spend for just the 21 SRM suppliers along with their associated ALC NSN-level spend. This includes more spend than the NSN-level spend in Figure 4.2 in Chapter Four. Figure B.1 shows how much more spend with the SRM suppliers is not included in SRM because the initiative is confined to spares and repair contracts. SRM suppliers were identified by the CAGE codes as provided by the GCSS-AF.[1] The darkened sections show the portion of total ALC spend that was used to identify AFMC's top sustainment suppliers. Two major types of spending are related to spares and repair items, identified by their NSNs: (1) spending on NSNs that are associated with CCs, split into spares and repair, and (2) spending for NSNs that are not associated with CCs, also split into spares and repair. An example of non-CC spending is aircraft-unique parts purchased by the ASWs.

The largest portion of ALC spending with SRM suppliers is not specific to any particular NSN. Goods and services purchased that are not specific to an NSN include contracts that do not have NSN Contract Line Item Numbers, such as services, Programmed Depot Maintenance (PDM), CLS, PBL, Total System Performance Responsiveness, modifications and upgrade contracts, and engineering and advisory and assistance service contracts. As we will note later, a large portion of the business that SRM suppliers have with AFMC, such as its product centers, is not subject to AFMC's current scope of SRM activities.

As noted earlier, AFMC identified its top sustainment suppliers for SRM by analyzing three years of NSN-level spend data (FY 2002 to 2004). Table B.1 shows the top 21 suppliers listed by their level of NSN-level spending.[2] According to several people we interviewed who participated in SRM from the beginning, AFMC's leadership drew the line at 21 suppliers to

---

[1] These CAGE codes were identified by the SRM teams and contractors. They represent those sites that provide sustainment goods and services for Air Force weapon systems. The downloaded file as of January 12, 2007, was entitled "SSRM Standard Report," which described FY 2002–2004 spend by CAGE code.

[2] Data for all budget program codes are included.

Figure B.1
SRM CAGE Code and Other ALC Spending ($22 Billion, Then-Year)

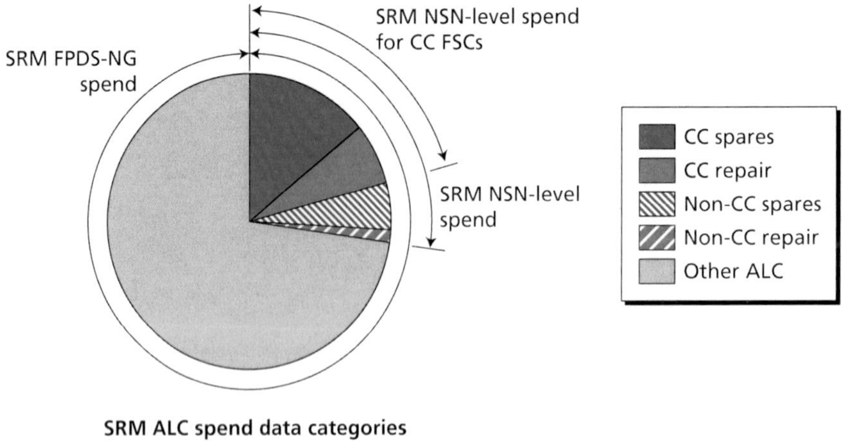

SOURCE: For SRM CAGE codes, FY 2006–2008 NSN-level spend (Strategic Sourcing Analysis Tool), FY 2006–2008 DD350 and FPDS-NG data (ALC purchase office codes).
RAND *TR904-B.1*

include several smaller or Tier I management companies.[3] The difference between the largest and smallest SRM supplier spend is over 100 to 1. Of interest was whether these smaller suppliers' issues—internally with their own companies and externally with the Air Force—differ from those of the prime contractors that have greater spend, such as the engine and aircraft manufacturers nearer to the top of the list. Total NSN-level spending summed to $6 billion for FY 2006–2008 in unadjusted dollars, as shown in Table B.1. This is the amount that most SRM teams focus on, though as noted in Figure B.1, this is less than the total spend at the three ALCs for these 21 suppliers.

In September 2005, BRAC required that reparable spares contracting move to DLA. Military service buyers would "transition in place," meaning that they would remain in their current locations but become DLA employees and report to DLA. Taking into account the fact that contracts for reparable spares would become a DLA responsibility, when we also analyzed repair spend separately, the need for SRM was still evident, as Figure B.2 shows. The organization of this figure is similar to that in Figure 4.3 in Chapter Four, in which we showed spending for repairs and spares associated with CCs at OC-ALC, OO-ALC, and WR-ALC from left to right. Suppliers are sorted in descending order of the extent that spend is distributed across the CCs.

Figure B.2 shows that SRM is especially important in the management and development of repair services with Honeywell, Smiths Aerospace (which has since merged with General Electric), Hamilton Sundstrand, Boeing, Goodrich, and DRS Technologies. Alternatively, near the bottom of the figure are those contractors that could be practically managed by a close collaboration between the SRM team and a single CC, though we note as before that this is largely due to the extent of spend being managed by SRM. These contractors include Kelly Aviation, EDO Corp, Raytheon, Pratt & Whitney, Teledyne, Crane Aerospace (Signal

[3] Here, *Tier I* refers to those companies that subcontract with the contractors holding primary contracts with the Air Force.

Table B.1
NSN Spending for 21 SRM Private-Sector Suppliers, FY 2006–2008

Supplier	$ (millions)
Pratt & Whitney	896
General Electric	754
Northrop Grumman	474
Boeing	446
Kelly Aviation	440
Honeywell	427
BAE Systems	386
Raytheon	377
Goodrich	345
Lockheed Martin	240
Rockwell Collins	192
EDO Corp.	174
L-3 Communications	139
Rolls-Royce Corp	114
Hamilton Sundstrand	114
Parker Hannifan	103
Teledyne	100
Smiths Aerospace	99
DRS Technologies	97
Chromalloy	31
Crane Aero (Signal Tech)	9
Total	5,957

SOURCE: FY2006–2008 NSN-spend for SRM CAGE codes, Strategic Sourcing Analysis Tool.

Technologies), and Chromalloy. The spend data showed no repairs for Rolls-Royce during the period of the study for these CC FSC components.

Because large contractors often do business with other military services and DLA, which may affect their incentives to respond to Air Force requests for cost and performance improvements, we next analyzed SRM spend with respect to total DoD spend. Assuming that contractors are most attuned to their larger customers, we reasoned that those contractors doing more business with the Air Force might be better candidates for expanding SRM to a larger spend pool.

In Figure B.3, the horizontal bars measure the percentage of business with spare, repair, other ALC, other AFMC, other Air Force, DLA SRM suppliers, other DLA, and other DoD,

Figure B.2
AFMC Spending by Firm and Commodity Council, Repair Only

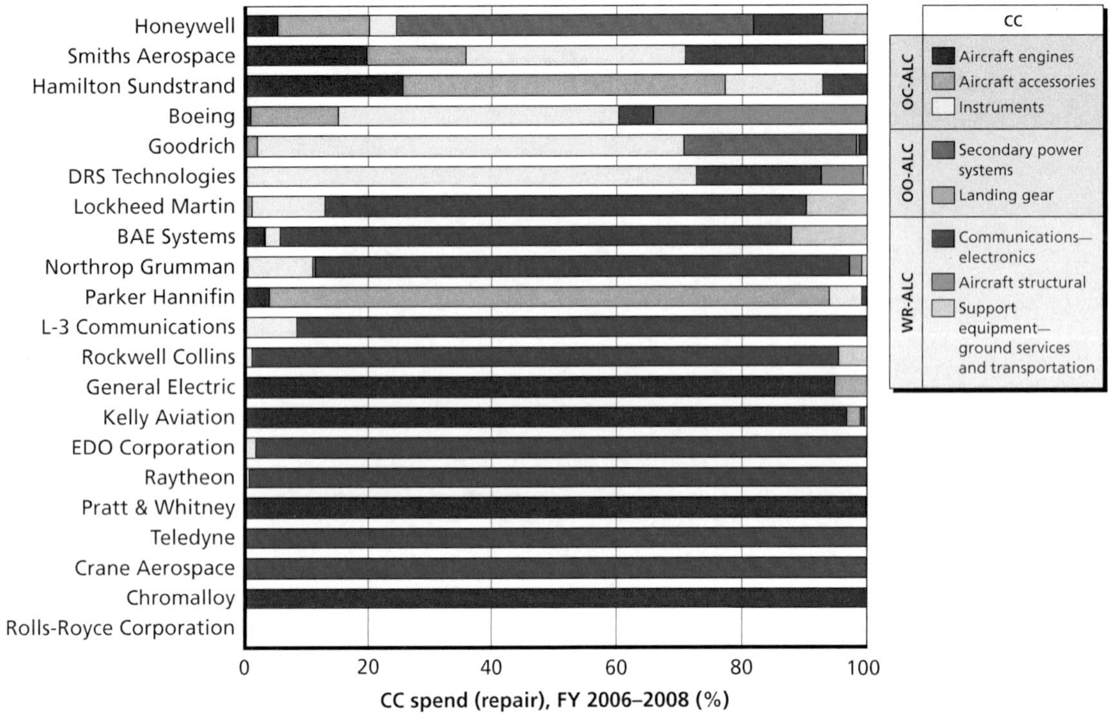

SOURCE: FY 2006–2008 NSN-level spend data for SRM CAGE codes and CC FSCs, Strategic Sourcing Analysis Tool.
RAND *TR904-B.2*

again moving from left to right.[4] In short, for these 21 suppliers the figure measures the percentage of its business for different parts of DoD. The right-hand side indicates the percentage of business that each contractor has with all of DoD. Thus, during the period of FY 2006–2008, 37 percent of all DoD contract dollars for the CAGE codes of the 21 SRM suppliers were obligated to the Lockheed Martin, 18 percent to Boeing, 10 percent to Northrop Grumman, and 9 percent to Raytheon. Pratt & Whitney came in at 4 percent and General Electric at 3 percent. The rest of the SRM suppliers range between 2 percent (Honeywell) and less than 1 percent.

When we look at Figure B.3, we note that the business of most of the Air Force's largest SRM suppliers largely derives from the Air Force. For example, almost half of Lockheed Martin's business for the SRM-related CAGE codes is with the Air Force, and over half of the CAGE code business of Boeing and Northrop Grumman is also with the Air Force. Of the large OEMs, only Raytheon represents much more direct business with services other than the Air Force.

Given the objective of SRM to bring about cost and performance improvements, the greatest financial incentives for both the Air Force and its SRM suppliers would seem to be to

[4] The spend data attributed to DLA is an overestimate of what is spent for the Air Force, because DLA also makes purchases for the same classes of items for the Army, Navy, and Marines. The data do not allow us to separate DLA spending by service.

Figure B.3
Key Supplier Spending with DoD

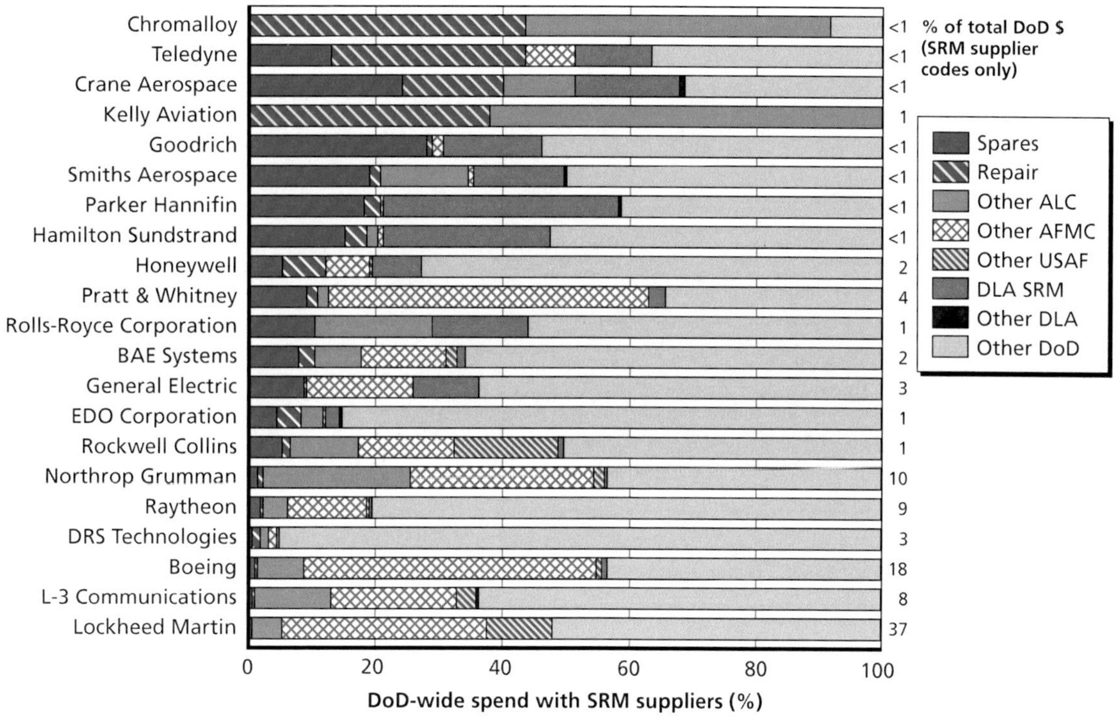

SOURCE: FY 2006–2008 DD350 and FPDS-NG data, NSN-level contract data (Strategic Sourcing Analysis Tool), and SRM CAGE codes.
NOTE: Percentages may not sum to 100 due to rounding.
RAND *TR904-B.3*

expand SRM to encompass more of the business base of these key suppliers. One way to move toward that direction would be to construct an Air Force supplier scorecard that encompasses all of the Air Force's business along the lines of best-in-class companies for best SRM practices.

APPENDIX C

Supplier Relationship Management Interviews

We conducted semi-structured interviews with SRM team personnel and collected material documentation of their work. We interviewed each of the five SRM teams as a group and most team members individually. The group interviews described the nature of the teams' work, while the individual interviews focused more on individual understanding of and experiences with SRM. Interviews were conducted in person or on the telephone (as available) and usually lasted one to two hours. They loosely followed a predetermined protocol (see Appendix B), allowing for interviewees to raise relevant topics not on the list. We followed up as needed with some team members when individual questions emerged. We took detailed notes during each interview, transcribing directly whenever possible. We analyzed the notes for consistent themes and for differences across the teams and across all SRM personnel.

Due to resource constraints, we were not able to interview all SRM stakeholders, such as SCMs, CCs, or suppliers. However, those whom we did interview were the persons actually charged with executing SRM, and they ought to have been most knowledgeable about the initiative and its requirements.

We verified the information that we collected from interviews with other material documentation and information from other interviews. We also relied on our extensive experience with qualitative data collection to construct, conduct, and analyze the interviews in such a way that we believe our interview data were fairly comprehensive and representative of the SRM team member's experiences during the study period.

APPENDIX D

Supplier Relationship Management Government and Private-Sector Company Interview Protocols

For this study, we conducted three categories of interviews: government SSRMs and their teams, individual Air Force SRM team members who were government civilian employees, and members of commercial firms that were best in class in SRM. For each category, we developed an interview protocol. These protocols were employed to guide the conversations, but they were not used as questionnaires. Frequently, topics were discussed that were not on the protocols but were relevant to the objectives of the study. All interviews were conducted on a volunteer basis.

Supplier Relationship Management (SRM) Protocol: Government SSRMs and SRM Teams

Background

RAND is conducting a study, entitled "Strategic Management of Maintenance," co-sponsored by Mr. Grover Dunn, AF/A4I, and Mr. Charlie Williams Jr., SAF/AQC. Its objective is to identify best practices—Air Force and commercial—in supplier management and development that have been shown to reduce total costs and improve performance of contract repair, supply, and services. The study will also highlight impediments to applying these best practices within the Air Force and to recommend ways of successfully adapting best practices to the Air Force context.

We appreciate your participation in our study and will keep individual identities confidential. Our results will be documented in briefings and a final report, which will be sent to you. Thank you for your time and participation.

I. Goals and Objectives

A. What is your vision for the SRM initiative that you are developing? What kind of direction are you trying to set?
B. Does your SRM initiative/team have additional goals beyond AFMC's PSCM Sustainment Transformation goals of increasing material availability, reducing costs, and reducing cycle time as measured by its Supply Chain Scorecard?
C. How are Sustainment Transformation and other goals and objectives communicated and reinforced to your team?
D. Do you use AFMC or DLA's Corporate Supplier Scorecard to measure your suppliers?

E. Are there plans to expand SRM beyond AFMC's top 21 suppliers and DLA and organic maintenance or to additional spend categories?

II. SRM Policy Guidance

A. Where do you receive your SRM policy guidance?
   1. Are you creating your own and if so, what guides you?
B. Do SSRMs coordinate with, learn from, and/or leverage each other's work? With SRM leaders in other organizations, such as DLA?
C. Are you aware of any SRM-related activities beyond AFMC in the Air Force or DoD?

III. Teams

A. How did you build your SRM team?
B. What is the current composition of your SRM team?
C. How are team members' time and efforts typically distributed?
D. What kinds of knowledge, skills, and experience are represented on your SRM team?
E. What kinds of SRM training or development have the SSRM and team members received?
F. How do team members stay current on best SRM practices?
G. How are your SRM resources funded/budgeted?
H. Have your suppliers developed *customer* teams to support SRM activities?
I. How is the performance of the SRM team and SRS personnel measured?
J. Is participation on the SRM team viewed as a career enhancement?

IV. SRM 4-Step Process

A. Are you developing and managing one supplier at a time, some more than others, or all suppliers simultaneously?
B. Where are you in the process with respect to each of your suppliers? Which of the following steps/activities have been done for each supplier?

Step 1: Assess the Current Environment
1. Develop Communication Infrastructure/Plan?
2. Perform SRS Staff Analysis?
3. Update and Publish Supplier Assessment Document (SAD)?

Step 2: Develop Supplier Strategy and Preliminary Objectives
1. Develop Supplier Strategy?
2. Brainstorm Preliminary Objectives (i.e., short-term improvement projects accomplished through JIIs)?
3. Develop Supplier Strategy Document (SSD)?

Step 3: Plan JIIS and Refine Objectives
1. Expand List of Objectives?
2. Analyze Feasibility and Impact of Objectives?

3. Update SDD?
4. Develop Supplier Initiative Management Plan (SIMP)?
5. Approve SIMPs?

Step 4: Execute JII Project Plans

1. Implement JII Project Plan?
2. Monitor & Report on Project Status?
3. Modify Project to Achieve Results?
4. Develop Initiative Closeout Report (ICR)?
5. Redefine Strategies?

V. Supplier Relationships and Supplier Development

A. What was the state of your supplier relationships when you first began SRM?
B. What are your supplier relationships like now?
   1. What are the most important factors for explaining any change, if any?
C. What would you like your supplier relationships to be?
   1. Are your suppliers (and team) aware of your expectations?
   2. Do you have a roadmap for getting to where you want them to be?
D. What makes achieving these future states particularly challenging?
E. What kinds of solutions would be helpful to helping you achieve these future supplier relationship states?

VI. Post-SRM Process Activities?

A. Is there feedback and continuous improvement in SRM processes?
B. What is being done to ensure improvements continue?
C. What kind of archival system is in place to document SRM efforts?

VII. Supplier Councils and Summits

A. What has been the value of the supplier councils?
   1. Are they providing what you need?
B. Are there plans to hold AFMC supplier summits separate from the AFMC/DLA summits?

VIII. Hindrances to Implementation of Best Practices in Supplier Management and Supplier Development

A. What are the top categories of formal and informal hindrances to SRM, if any, and why? For example (though not limited to),
   1. Data availability, quality, and timeliness?
   2. Personnel skills needed for advancing joint improvement efforts?
   3. Existing long-term contracts?
   4. Limited leverage with suppliers?

5. Policies (federal, DoD, AF)?
6. Resistance from suppliers?

What have we not asked about SRM that you think is important?

Supplier Relationship Management (SRM) Protocol: Government Civilian Employees in SRM

Background

RAND is conducting a study, entitled *Strategic Management of Maintenance*, co-sponsored by Mr. Grover Dunn, AF/A4I, and Mr. Charlie Williams Jr., SAF/AQC. Its objective is to identify best practices—Air Force and commercial—in supplier management and development that have been shown to reduce total costs and improve performance of contract repair, supply and services. The study will also highlight impediments to applying these best practices within the Air Force and to recommend ways of successfully adapting best practices to the Air Force context.

We appreciate your participation in our study and will keep individual identities confidential. Your answers will be combined with answers from other SRM teams and analyzed as a group, not individually. Our results will be documented in briefings and a final report, and none of the answers will be able to be traced to you. We will send a copy of the final report to you. Your participation in this interview is completely voluntary, and you will not be penalized if you choose not to participate or not to answer any individual questions. However, your input is very important to us and this study, and we hope you will choose to participate. Are you willing to be interviewed?

Thank you for your time and participation.

I. Current Position and Background

A. Tell me about your current position; what is your career series?
   1. What do you do on a typical day?
   2. Does this change at different times of the year?
   3. Where does SRM fall in the pecking order of your various responsibilities?
B. How long have you been in this position? How were you recruited?
C. What did you do before you came to this position?
D. What are the skills or qualities that you most need for this position? Were they what you thought you would need before you worked in this position?

II. Supplier Relationship Management

A. Background and Training

1. When did you first hear people talking about SRM?
2. Do you believe SRM is helping the Air Force?
3. Can you describe the SRM training that you have received? (Content, frequency, quality, when, how often)

B. Use of SRM

1. Are you on an SRM team?
   - If yes, who else is on your team—please explain their roles.
2. Can you describe AFMC's relationships with your suppliers?
   - Have these relationships changed since SRM?
3. Can you describe your own personal relationships with your suppliers?
   - Have these relationships changed since SRM?
   - Who in the suppliers' organization do you communicate with the most?
   - What has been most successful?
   - What has been most surprising?
4. Has SRM changed your day-to-day activities at all? How?
5. Have these changes been useful at all? Has SRM made your job easier or harder? Please explain.
6. Can you explain your use of the following and tell me how useful or successful these SRM components are. (Omit those already discussed.)
   a. Joint Improvement Initiatives
   b. Supplier councils/summits
   c. Supplier scorecards
   d. Collaborative forecasting
   e. Supplier integration
   f. SRM offsites (e.g., Tier III meetings)
   g. Others (please list)
7. How about other people on the SRM team—do they seem to really use SRM? If yes, how? If no, why not?
8. How about contracted personnel in your office—how do they help or not help the SRM team?

C. Improving SRM

1. What are some of the challenges or problems you have encountered with implementing SRM?
2. What do you wish your superiors understood about your job that SRM does not account for?
3. What do you wish suppliers understood about your job that SRM does not account for?
4. What would help you (in SRM or otherwise) to complete your SRM responsibilities more successfully?
5. Do you feel that you are making a difference with SRM?
6. Beyond your position, do you think that SRM benefits the Air Force? How?
7. Do you think SRM will stay around or not? Please explain.
8. What else should we know about SRM that we have not asked?

III. Hiring/Advancement/Qualifications

A. SRM

1. Is SRM team experience considered an asset in your career series?
2. What is considered an asset?
3. How do you know whether you and your team are doing well?
4. What are people doing now who did this job before?

B. Advancement

1. How do people advance in your career series? What criteria help people get ahead?
2. Are these the same criteria that are outlined in the [Quality Ranking Criteria/Civilian Force Development/Whole Person Score]?
3. Does SRM help you gain any of these skills or criteria that are needed for advancement?
4. Are there skills or qualifications you need for SRM that are not taken into account by the [Quality Ranking Criteria/Civilian Force Development/Whole Person Score]?

Supplier Relationship Management (SRM) Protocol: Commercial

[Company Name]

Background

RAND is conducting a study, entitled *Strategic Management of Maintenance*, co-sponsored by Mr. Grover Dunn, Director of Transformation, Air Force Deputy Chief of Staff for Logistics, Installations, and Mission Support (AF/A4I), and Mr. Charlie Williams Jr., Deputy Assistant Secretary for Contracting, Office of the Assistant Secretary of the Air Force for Acquisition (SAF/AQC).

The objective of this study is to identify best practices—Air Force and commercial—in supplier management and development that have been shown to reduce total costs and improve performance of contract repair, supply and services. It will also highlight impediments to applying these best practices within the Air Force and recommend ways of successfully adapting best practices to the Air Force context.

Protocol Questions

1. Background
   a. When and why was your SRM program created?
   b. About how many (or of what type) of your suppliers are included in your SRM efforts and how were they identified?
   c. Where are your SRM efforts located in your organization? Has it changed over time?
   d. Who are the key stakeholders that you have needed to "buy in" to SRM and how have you achieved that buy in?

2. What are the specific goals of your supplier development program?

3. How do you develop your SRM teams?
   a. What kinds of training do you provide?
   b. What kinds of skills do members have?
   c. Do you develop suppliers simultaneously for aftermarket support and production?

4. Supplier Development Perspectives
   a. About what proportion of organization's SRM efforts are spent on reactive versus proactive activities?
   b. Do you have formal SRM plans for each supplier you are developing?
   c. Do you have suppliers that are not fully engaged with your SRM efforts? If so, how do you plan to get them more fully engaged?

5. What are the critical success factors in your SRM efforts? For example,
   a. Is your SRM perspective enterprise-wide, business unit–wide, or some other scope?
   b. Does top management support your SRM efforts?
   c. Do your SRM efforts have cross-functional support?
   d. Are your SRM analyses performed enterprise-wide, business-wide, or some other scope?
   e. Is Total Quality Management (TQM) a part of SRM?
   f. Are you continually refining your SRM efforts?

6. Do you have a Supplier Development Model? If so, could you please describe it?

7. Supplier Scorecards
   a. What kinds of metrics (generally speaking) are used in your supplier scorecard?
   b. How do you use the scorecard in your SRM activities?
   c. Are supplier scorecard values visible to all your suppliers?
   d. Is the supplier scorecard used for all of the supplier's business?

8. Supplier Summits
   a. Do you hold Supplier Summits?
   b. If so, how are they organized?
   c. Do the summits lead to actionable items? How do you ensure that concrete constructive steps results from them?

9. What are the most important investments you have made in your SRM efforts?
   a. For example, information technology, training, number and type of team members, time and resources spent with suppliers, etc.
   b. Did you develop business cases to justify these investments?

10. Assessing the Benefits of SRM
    a. What, if any, impacts from SRM have you observed and how have you attributed it to SRM?
    b. How do you assess the benefits of your SRM efforts?
    c. How do you justify the resources allocated to your SRM efforts?

11. What are the key lessons you would recommend the Air Force take from [company name's] experience in SRM to apply to its own initiative to develop and manage its top suppliers?

Bibliography

Air Force Directorate on Integration and Transformation (AF/ILI) eLog21, "Commodity Councils—Fact Sheet," AFD-060831-049, August 31, 2006. As of October 4, 2010:
http://www.af.mil/shared/media/document/AFD-060831-049.pdf

AFMC Public Affairs, "Command Moves Forward with Restructuring Plan," May 11, 2010. As of October 4, 2010:
http://www.afmc.af.mil/news/story.asp?id=123204051

AFMC Supplier Relationship Management Guide, Version 2, Air Force Materiel Command, Wright-Patterson Air Force Base, June 2006.

Amaral, Jason, Corey A. Billington, and Andy A. Tsay, "Outsourcing Production Without Losing Control," *Supply Chain Management Review*, November/December 2004, pp. 51–52.

Avery, Susan, "Cessna Soars," *Purchasing*, September 4, 2003.

———, "Lean, but Not Mean, Rockwell Collins Excels," *Purchasing*, September 1, 2005.

Boeing, "Supplier Performance Measurement," no date. As of December 3, 2007:
http://www.boeing.com/companyoffices/doingbiz/supplier_portal/index_general.html

Bowersox, John J., David J. Closs, and M. Bixby Cooper, *Supply Chain Logistics Management*, New York: McGraw-Hill, 2002.

Bowman, Robert, "For Boeing, a New Aircraft Means a Revamped Supply Chain," *Global Logistics & Supply Chain Strategies*, March 1, 2007.

CAPS Research, *Purchasing Performance Benchmarking Report for the Aerospace/Defense Industry*, Tempe, Ariz., January 24, 2005.

———, *Aerospace and Defense Industry 2007 Supply Management Performance Benchmarking Report*, Tempe, Ariz., May 18, 2007.

CAPS Research, A. T. Kearney, and the Institute for Supply Management, "Succeeding in a Dynamic World: Supply Management in the Decade Ahead," *Inside Supply Management*, May 2007.

Carbone, James, "Altel Slashes Cost with Strategic Sourcing," *Purchasing*, July 14, 2007.

———, "Sun Microsystems Drives Suppliers to Success," *Purchasing*, October 20, 2005a.

———, "The Changing Face of Nortel," *Purchasing*, May 19, 2005b.

———, "Motorola Leverages Its Way to Lower Costs," *Purchasing*, September 16, 2004a.

———, "Hewlett-Packard Wins for the 2nd Time," *Purchasing*, September 2, 2004b.

———, "Alcatel Buyers Focus on Cost, Time-to-Market," *Purchasing*, March 18, 2004c.

———, "Motorola Simplifies to Lower Costs," *Purchasing*, October 18, 2001.

———, "Reinventing Purchasing Wins the Medal for BIG BLUE," *Purchasing*, September 16, 1999.

Carter, Phillip L., et al., *Succeeding in a Dynamic World: Supply Management in the Decade Ahead*, Institute for Supply Management and W. P. Carey School of Business at Arizona State University, 2007.

Cessna, "Integrated Supply Chain Process," brochure, 2007.

Chapman, Timothy L., Jack J. Dempsey, Glenn Ramsdell, and Michael R. Reopel, "Purchasing & Supply Management: No Time for Lone Rangers," *Supply Chain Management Review*, Winter 1998.

Ellram, Lisa M., *Strategic Cost Management in the Supply Chain: A Purchasing and Supply Management Perspective*, Tempe, Ariz.: CAPS Research, August 2002.

Ericksen, Paul D., and Rajan Suri, "Managing the Extended Enterprise," *Purchasing Today*, Vol. 12, No. 2, February 2001, p. 58.

Federal Procurement Data System, homepage, no date. As of October 7, 2010:
https://www.fpds.gov

Feller, Andrew, Dan Shunk, and Tom Callerman, "Value Chains Versus Supply Chains," *Business Process Trends (BPTrends)*, March 2006. As of October 4, 2010:
http://www.value-chain.org/en/art/?120

Fitzgerald, Kevin R., "For Superb Supplier Development—Honda Wins!" *Purchasing*, September 21, 1995.

GAO—*See* U.S. General Accounting Office.

Goodman, Russell, "HP Shows How to Outsource Production While Keeping Control Over Suppliers," SupplyChainBrain, December 1, 2006. As of October 4, 2010:
http://www.supplychainbrain.com/content/technology-solutions/edi-communication-xmledi/single-article-page/article/hp-shows-how-to-outsource-production-while-keeping-control-over-suppliers/

Headquarters, Air Force Materiel Command, *Supply Chain Scorecard, Key Performance Indicators and Definitions*, Wright-Patterson Air Force Base, September 15, 2006.

Honda, "How to Become a Supplier," no date. As of October 4, 2010:
http://www.hondasupplyteam.com/j_pstat/html/honda_howto_supplier.htm

Johnson, Clay III, Deputy Director for Management, Office of Management and Budget, "Implementing Strategic Sourcing," memorandum, May 20, 2005. As of October 4, 2010:
http://www.uspto.gov/web/offices/ac/comp/proc/OMBmemo.pdf

Jones, Daniel T., and James P. Womack, *Seeing the Whole: Mapping the Extended Value Stream*, Cambridge, Mass.: Lean Enterprise Institute, 2002.

Kotter, John P., and Dan S. Cohen, eds., *The Heart of Change: Real-Life Stories of How People Change Their Organization*, Boston: Harvard Business School Press, 2002.

Krause, Daniel R., and Robert B. Handfield, *Developing a World-Class Supply Base*, Tempe, Ariz.: Center for Advanced Purchasing Studies, 1999.

Laseter, Timothy M., "Balanced Sourcing the Honda Way," *Strategy + Business*, Fourth Quarter 1998.

Lee, Hau L., "The New Frontier of Supply Management," *Inside Supply Management*, Vol. 13, No. 2, February 2002, p. 4.

Leenders, Michael, and Harold E. Fearon, *Purchasing and Supply Management*, 11th edition, Chicago: Irwin, 1997, as reprinted in John J. Bowersox, David J. Closs, and M. Bixby Cooper, *Supply Chain Logistics Management*, New York: McGraw-Hill, 2002.

Licker, Jeffrey K., *The Toyota Way: 14 Management Principles from the World's Greatest Manufacturer*, New York: McGraw-Hill, 2004.

Licker, Jeffrey K., and Yen-Chun Wu, "Japanese Automakers, U.S. Suppliers and Supply-Chain Superiority," *Sloan Management Review*, Fall 2000.

Lundsford, Lynn, and Daniel Michaels, "More Delays Plague Boeing's Dreamliner," *Wall Street Journal*, January 16, 2008.

Monczka, Robert M., and William J. Markham, "The Future of Supply Management: Part 1—Category Strategies and Supplier Management," *Supply Chain Management Review*, September 2007.

Monczka, Robert M., Robert J. Trent, and Robert B. Handfield, *Purchasing and Supply Chain Management*, 2nd edition, Cincinnati, Ohio: South-Western College Publishing, 2002.

Moore, Nancy Y., Laura H. Baldwin, Frank Camm, and Cynthia R. Cook, *Implementing Best Purchasing and Supply Management Practices: Lessons from Innovative Commercial Firms*, Santa Monica, Calif.: RAND Corporation, DB-334-AF, 2002. As of October 4, 2010:
http://www.rand.org/pubs/documented_briefings/DB334.html

Morgan, James P., "Cessna Charts a Supply Chain Flight Strategy," *Purchasing*, September 7, 2000, pp. 42–61.

Muller, Joann, "A Savior from the East," *Forbes*, June 4, 2007, pp. 112–113.

Nelson, Dave, Rick Mayo, and Patricia E. Moody, *Powered by Honda: Developing Excellence in the Global Enterprise*, New York: John Wiley and Sons, 1998.

Nelson, Dave, Patricia E. Moody, and Jonathan R. Stegner, *The Purchasing Machine*, New York: The Free Press, 2001.

———, *The Incredible Payback: Innovative Sourcing Solutions That Deliver Extraordinary Results*, New York: American Management Association, 2005.

Office of the Under Secretary of Defense (Comptroller), *National Defense Budget Estimates for FY 2010*, June 2009. As of October 4, 2010:
http://comptroller.defense.gov/defbudget/fy2010/Green_Book_Final.pdf

Ohnsman, Alan, "Toyota, Honda Say Quake Won't Disrupt U.S. Shipments (Update 1)," *Bloomberg*, July 19, 2007. As of October 4, 2010:
http://www.bloomberg.com/apps/news?pid=20601101&sid=a8CCY9eaHeBg&refer=japan#

Pazmany, Peter, "Creating the Extended Enterprise: Sun's Virtual Network," *Supply Chain Management Review*, November/December 2000, p. 56.

Pfeiffer, Matt, AFMC/A4KS, "Top 21 KPI Drill-Down Analysis SSRM Offsite 31 Jan 06," briefing, Wright-Patterson Air Force Base, January 31, 2006.

Roche, James G., and Gen John P. Jumper, "Expeditionary Logistics for the 21st Century Campaign Plan," AFD-060831-041, August 31, 2006. As of February 8, 2010:
http://www.af.mil/shared/media/document/AFD-060831-041.pdf

Siekman, Philip, "GE Bets Big on Jet Engines," *Fortune*, December 19, 2002.

Stegner, John, "Gloves on the Boardroom Table," in John P. Kotter and Dan S. Cohen, eds., *The Heart of Change: Real-Life Stories of How People Change Their Organization*, Boston: Harvard Business School Press, 2002.

Stundza, Tom, "New Era Dawns at Boeing," *Purchasing*, August 10, 2000.

Teague, Paul E., "Product Development: Material and Supply Joins Design Teams," *Purchasing*, September 1, 2005.

———, "Purchasing Names Boeing Executive Supply Chain Manager of the Year," *Purchasing*, October 18, 2007.

Trent, Robert J., and Robert M. Monczka, "Purchasing and Supply Management Trends and Challenges Throughout the 1990s," *International Journal of Purchasing and Materials Management*, Fall 1998, pp. 3–4.

U.S. Air Force, *Expeditionary Logistics for the 21st Century*, Washington, D.C., April 1, 2005.

U.S. Department of Defense, Defense Manpower Data Center, Statistical Information Analysis Division, "DoD Personnel & Procurement Statistics: Personnel & Procurement Reports and Data Files," website, no date. As of October 8, 2010:
http://siadapp.dmdc.osd.mil/index.html

———, "DoD Procurement: Procurement Reports and Data Files for Download," web page, 2008. As of October 8, 2010:
http://siadapp.dmdc.osd.mil/procurement/Procurement.html

U.S. General Accounting Office, *Depot Maintenance: DoD's 50-50 Reporting Should Be Streamlined*, Report to Congressional Committees, GAO-03-1023, September 2003. As of October 4, 2010:
http://www.gao.gov/new.items/d031023.pdf

Warren, David R., *Depot Maintenance: Air Force Faces Challenges in Managing to 50-50 Ceiling*, Testimony Before the Subcommittee on Readiness, Committee on Armed Services, U.S. Senate, GAO-T/NSIAD-00-112, March 3, 2000. As of October 4, 2010:
http://www.gao.gov/archive/2000/ns00112t.pdf

Wikipedia, "Six Sigma," no date. As of October 11, 2010:
http://en.wikipedia.org/wiki/Six_sigma

Womack, James P., and Daniel T. Jones, *Lean Thinking: Banish Waste and Create Wealth in Your Corporation*, New York: Simon and Schuster, 1996.